SITING HAZARDOUS WASTE FACILITIES

Local Opposition and the Myth of Preemption

DAVID MORELL
and
CHRISTOPHER MAGORIAN
Center for Energy and Environmental Studies
Princeton University

BALLINGER PUBLISHING COMPANY
Cambridge, Massachusetts
A Subsidiary of Harper & Row, Publishers, Inc.

International Standard Book Number: 0-88410-906-2

Library of Congress Catalog Card Number: 82-1789

Printed in the United States of America

Library of Congress Cataloging in Publication Data

Morell, David.
Siting hazardous waste facilities.

Bibliography: p.
Includes index.
1. Waste disposal sites—United States. 2. Chemical industries—United States—Waste disposal. I. Magorian, Christopher. II. Title.
TD811.5.M67 363.7'28 82-1789
ISBN 0-88410-906-2 AACR2

SITING HAZARDOUS WASTE FACILITIES

DEDICATION

For Andrea

. . . .

D.M.

For My Parents

. . . .

C.M.

CONTENTS

Appendices

LIST OF FIGURES

ACKNOWLEDGMENTS

The original research on which this book has been based was conducted at Princeton University's Center for Energy and Environmental Studies under a grant from the Andrew W. Mellon Foundation. This research on siting of hazardous waste management facilities was one component of a larger project at Princeton on management of hazardous wastes which the Mellon Foundation has made possible. Supplemental support for the siting research was received from the Office of Science and Technology Policy, Executive Office of the President.

We wish to express our thanks to the many individuals and groups who gave so willingly of their time and their insights during the course of this research, and in responding to the final pre-publication version of this manuscript. The critique from Michael O'Hare of Harvard University was especially comprehensive and insightful, as were the thoughts of Diane Graves of the New Jersey Sierra Club. We also appreciated the opportunity to review the pre-publication draft of the excellent manuscript, *Facility Siting,* by O'Hare, Debra Sanderson, and Lawrence Bacow. Others to whom we owe a special note of appreciation include: Paul Abernathy of Chemical Waste Management, Inc.; Richard Gimello of the New Jersey Department of Environmental Protection; David Goetze of Resources for the Future; Curtis Haymore of the United States Environmental Protection Agency; Michael

Italiano of the Office of Science and Technology Policy; Anne Kruger of the Center for Energy and Environmental Studies; David Moldenhauer of Princeton University; Kent Stoddard of the California Office of Appropriate Technology; and Lawrence Susskind of the Massachusetts Institute of Technology. They, among others, helped us comprehend the intricacies of the siting process. For any errors of fact or judgment which remain, we, rather than they, are responsible.

The opinions and conclusions presented in this book are those of the two authors and not necessarily those of the Mellon Foundation, the Office of Science and Technology Policy, or Princeton University.

David Morell
Christopher Magorian
Princeton, N.J.
February 1982

SITING HAZARDOUS WASTE FACILITIES

1 THE HAZARDOUS WASTE CRISIS AND THE SITING IMPERATIVE

More effort has gone into regulation of restaurants and taxicabs than into establishing a safe network for waste disposal.

Douglas Costle and Eckhardt Beck, EPA officials (1980)[1]

There's going to be a lot of screaming; more screaming and hollering than you'll ever see in your life in this county.

Stephen Capestro, New Jersey County Freeholder (1980)[2]

THE NATURE OF THE CRISIS

The issue of how to dispose properly of hazardous wastes has only recently gained the notoriety and alarm it deserves. Environmental experts now believe that problems of hazardous waste management will be, in the words of former Environmental Protection Agency (EPA) Administrator Douglas Costle, ". . . our single greatest environmental challenge in the next couple of decades."[3] These wastes pose an enormous danger. Though they have diverse chemical properties and vary widely as to the precise degree of hazard, overall such wastes present a danger of environmental destruction and of harm to human health. Perhaps the biggest threat from chemical wastes lies in the possibility of groundwater contamination, an especially serious problem since groundwater accounts for approximately half of the nation's drinking water supplies.[4]

Aside from the physical dangers of the wastes themselves, attempts to site hazardous waste management facilities are accompanied by grave political risks as well. Unlike most public goods, a proposed new hazardous waste facility frequently is cast by the host community as a matter of life and death. Local responses to siting proposals—especially for landfills—are often not only contentious but outright explosive. Attempts to impose these facilities on unwilling areas thus represent one of the most difficult challenges to a society which accepts the need for governmental efforts to promote the collective well-being, yet which also stresses the inherent rights of individuals to determine their own destiny.

The political and institutional difficulties of proceeding with construction of a proposed facility at a site that seems, initially, to meet appropriate engineering and technical criteria time and again bring the siting process to a halt. The fundamental difficulty is not one of finding a location that can meet engineering criteria acceptably, but of obtaining local consent and the state or federal permits necessary to build the proposed new facility at that location.[5]

The siting of hazardous waste facilities, then, involves basic political dilemmas, among the most important being the conflict between majority rule and minority rights. In fact, the decisionmaking process which Dennis Ducsik has characterized as "decide-announce-defend"[6] itself breeds conflict rather than consensus. Mechanisms to incorporate people's true feelings are not adequately incorporated into siting decisions. Conflicts are exacerbated rather than resolved amicably. And alienation of local citizens grows rather than diminishes.

Communities typically respond to plans to build a hazardous waste facility with the view: "Not in *my* backyard."[7] Whatever the details of the actual facility being proposed, waste facility siting proposals appear to the public as another landfill or a "dump," to be opposed strenuously. A national public opinion survey conducted in 1980 for the Council on Environmental Quality (CEQ), for example, found that people objected greatly to having a hazardous waste facility sited in their area. CEQ's researchers concluded that: "Aversion to the siting of a disposal site for hazardous waste chemicals is very strong despite the fact that respondents were given special assurances." Even when respondents were told that "disposal could be done safely and that the site would be inspected regularly for possible problems," this kind of facility was endorsed by a majority of respondents only when its distance from their own homes exceeded 100 miles.[8]

Local opposition to proposed hazardous waste management facilities—the "not in my backyard" reactions—may appear parochial and uninformed. At times, ignorance may prevail. Often, however, fears about the facility's potential dangers to public health in the area and about its potential impacts on the community's image are highly rational. As Michael O'Hare, Debra Sanderson, and Lawrence Bacow have concluded about facility siting controversies in general: "Far from finding the participants irrational, we are struck by the consistency with which parties in facility siting disputes act rationally and effectively to serve their interests as they perceive them."[9]

Whatever its particular causes, local opposition is a principal barrier to obtaining sites for new waste treatment, storage, and disposal facilities. And such opposition can eventually undermine the national regulatory process for hazardous wastes now underway under the Resource Conservation and Recovery Act (RCRA) of 1976. As a case-study report by Centaur Associates to EPA in 1979 concluded: "Public opposition to siting of hazardous waste management facilities, particularly landfills, is . . . the most critical problem in developing new facilities."[10]

This typical public response to proposed new waste facilities signals a challenge to the political process to devise a satisfactory means to reconcile the legitimate claims of those opposing a new site with the broader goal of establishing some of these necessary new facilities. The first reaction to this challenge has been a search for easy answers.

Simple solutions to the problem of siting new hazardous waste facilities have an appeal associated with their very simplicity.

> Let the states decide on their own. . . . The locals will always say "no" anyway.
>
> Put all these facilities out in the desert; then no one will care.
>
> Locate them all on federal land (or military reservations), where local (and even state) governments have no jurisdiction.[a]
>
> Let's get on with the job, using eminent domain to acquire all the sites we need.

Unfortunately, the misleading veneer of such proposals obscures the emotions and legitimate concern associated with local opposition to proposed new hazardous waste facilities.

[a]Such sites would pose especially difficult problems for inspection, monitoring, and enforcement of state regulations.

Concern over impediments to siting is by no means purely academic, of course. Indeed, the process of siting new hazardous waste management facilities poses extraordinary difficulties in achieving a balance between the nation's interest in public health and safety, and the public's demand for products of the chemical industry and for other consumer products the manufacture of which generates dangerous wastes: medicine, drugs, insulation, metals, and so on. Figure 1–1 shows the seven industries which accounted for approximately sixty percent of all industrial hazardous wastes in the United States in 1975. EPA has estimated that about ten percent of all industrial wastes pose a possible threat to human health or living organisms. These include toxic chemicals, pesticides, heavy metals, oils, acids,

Figure 1–1. Hazardous Wastes Generated by Selected Industries, 1975.

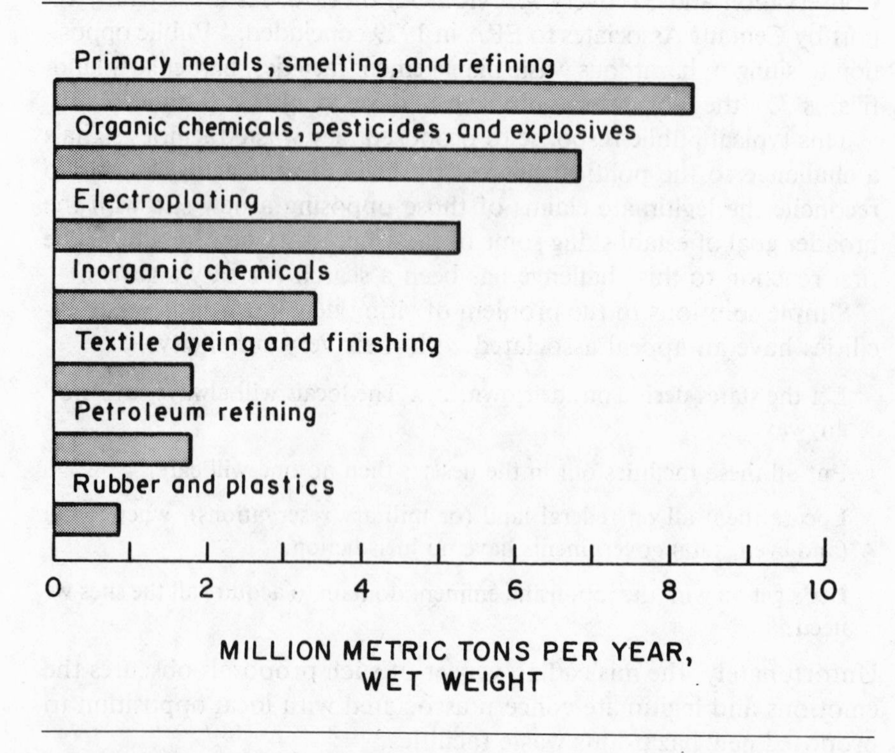

Source: Council on Environmental Quality et al., *Environmental Trends* (Washington, D.C.: U.S. Government Printing Office, 1981), p. 85.

caustics, flammables, explosives, and radioactive wastes. Most are liquids or semiliquids (sludge), but some are solids or gases. They are created by industries, farms, government installations, hospitals, and laboratories.[11]

The magnitude of the problem is staggering. EPA has estimated that 43 million metric tons of nonradioactive hazardous wastes were generated in the United States in 1981—approximately one pound per person per day—and that this level of waste generation can be expected to increase by about three percent each year.[12] New sites and new facilities are thus needed to treat, store, and dispose of all these new wastes. While this need exists nationally, hazardous wastes are concentrated in certain regions of the country—those where industry is also concentrated. Figure 1–2 shows industrial hazardous waste generation in various regions in 1975. Region V, in the Midwest, accounted for 25 percent of the national total. This region has the largest amounts of hazardous wastes from six industries: batteries, primary metals, electroplating, special machinery, paints, and waste oil from refining. Region VI, which includes the giant Texas-Louisiana petrochemical complex, accounted for another 25 percent of the national total. This area ranked first in wastes associated with inorganic and organic chemicals, explosives and pesticides, and petroleum refining.[13] The eleven states in these two regions were responsible for half of all the hazardous wastes coming from the entire American economy.

The United States also needs new sites and new facilities in which to place those wastes which still remain to be removed from the terrible legacies of past improper disposal: Love Canal (New York), Valley of the Drums (Kentucky), Price's Pit (New Jersey), and all the other areas whose severe impact on public health is becoming more evident daily. In 1979, a study by Fred C. Hart Associates for EPA estimated that between 32,000 and 50,000 hazardous waste disposal sites exist nationally. In the near future as many as 1,200 to 2,000 of these sites may pose significant dangers to health or the environment.[14]

Most potentially hazardous wastes have been disposed of on the land. Common disposal methods have included placing the wastes in unlined lagoons and similar surface impoundments (50 percent); placing them in dumps or other landfills (30 percent); disposing of the wastes through uncontrolled burning (10 percent); and deep well injection of wastes (less than 10 percent).[15] Cleanup of many of these old sites is essential if the public's health is to be protected—but where

Figure 1-2. Industrial Hazardous Wastes Generated by Region, 1975.

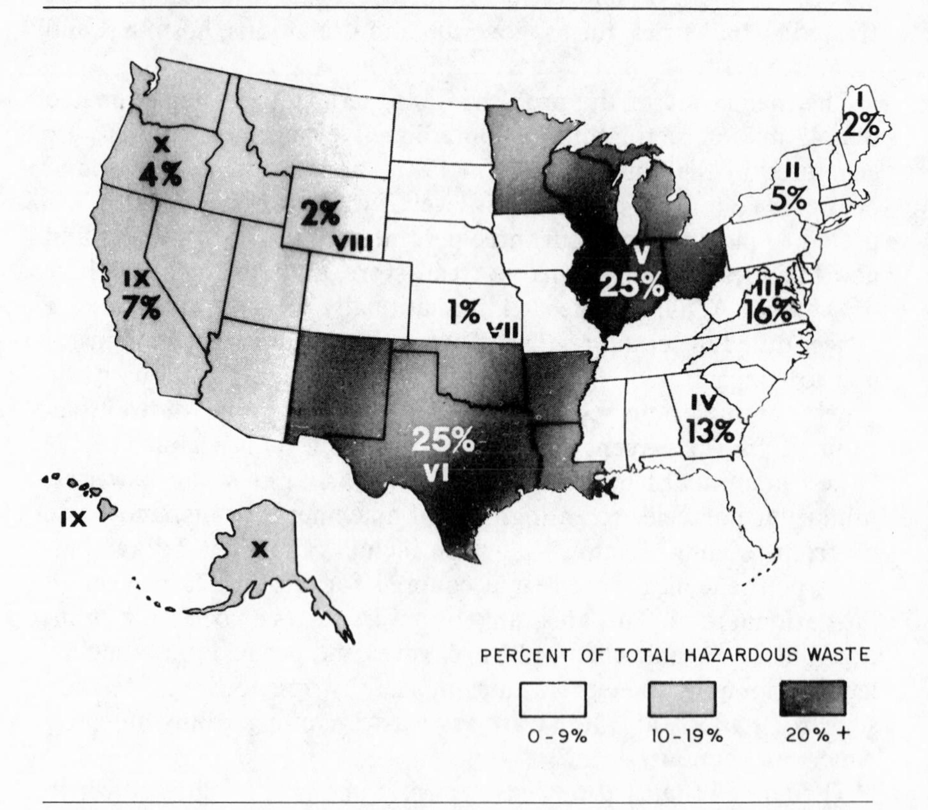

Source: Council on Environmental Quality et al., *Environmental Trends* (Washington, D.C.: U.S. Government Printing Office, 1981), p. 85.

are these wastes supposed to go? Again, new facilities will be required. In sum, as a group of experts from government, industry, and citizen groups at a 1980 meeting sponsored by the Keystone Center in Keystone, Colorado, concluded: "Establishing procedures and practices to assure the safe handling of non-radioactive hazardous wastes may be the most important public health issue in the United States during the 1980s."[16]

Such facilities, while needed, will not be easy to site. Given the growing record of leaking landfills and contaminated water,[17] resistance to facility siting proposals—especially for new landfills—by citizens and local governments is wholly understandable, and readily predictable. The siting problem thus may be seen in its starkest dimensions. For while sites for new facilities are needed by society, most siting proposals will be opposed strenuously by the residents of the particular area which has been chosen to host the facility. When the imperatives of the broader society (the majority) come into direct confrontation with the interests of one of its components (the minority), political tensions can become explosive. The Keystone conferees were of the opinion that:

> development of new sites for hazardous waste management facilities will take place in a climate of heightened public concern and low credibility of government and industry. . . . [There is] deep concern in local communities about proposals to build new hazardous waste management facilities.[18]

Not every site proposed by industry or government for a new hazardous waste facility deserves to be approved, of course. Effective public participation through an acceptable siting process can help identify precisely those sites and facilities which should be rejected.

To many citizens, proposals by industry or government to locate a new hazardous waste facility in their community immediately conjure up images of Love Canal in Niagara Falls, New York, where a covered-over landfill of Hooker Chemical Company wastes from production that occurred years earlier caused serious health problems for residents whose houses were built near the abandoned dump. Seven hundred and ten people were relocated following such actions as President Jimmy Carter's August 1978 declaration of a national emergency.[19] Love Canal, of course, is only the most famous of the many tragedies already familiar to a cynical but frightened populace; Appendix A examines the dynamics of a similar episode in Jackson Township, New Jersey.

Though handling of hazardous wastes in the past has been grossly inadequate, the federal government and many states in recent years have begun to exercise greater control over disposal of these dangerous products. The 1976 Resource Conservation and Recovery Act (RCRA) forms the centerpiece for the nation's hazardous waste management program. This act's principal provisions entail: definition of hazardous wastes; a manifest system to track hazardous wastes as they are transported;[20] standards for generators and transporters of hazardous waste; permit requirements for treatment, storage, and disposal facilities; and requirements for state programs.[21] Although

EPA failed to promulgate its regulations to enforce RCRA until several years after its passage, most of them went into effect on November 19, 1980. By late 1981, many states were seeking federal approval of their own programs. Some had devised their own state regulations to implement RCRA; these state regulations may be more stringent than the federal ones.[22] There was, however, further delay in EPA issuance of several important regulations, including those setting standards for land disposal of hazardous wastes.[23] As a result of inadequate enforcement in the past, the proclamations of federal and state regulators frequently fall on suspicious if not cynical ears.

Hazardous waste disposal and treatment operations assume a variety of forms. The most familiar—and feared—is land disposal, in modern terminology a "secure landfill." These facilities differ from general purpose or sanitary landfills:

> Secure landfills are designed to prevent connection with ground and surface waters and to prevent different wastes from coming into contact with each other. This is usually accomplished with liner and capping materials (e.g., clay and synthetic liners), separate cells for specific waste types, continuous monitoring, and leachate collection system.[24]

Recent waste management practices dictate that reliance on landfills be strictly limited to those materials which cannot possibly be recycled or treated in some other manner. New Jersey and California, for example, have placed sharp new constraints on further landfilling of hazardous wastes in their states. RCRA regulations constrain disposal of liquid wastes in landfills, a practice typical of the past.

Use of more sophisticated waste treatment and incineration technologies has expanded in the last few years, and growing public pressure is reinforcing state moves in this direction. California's Office of Appropriate Technology, for one, has studied use of these techniques and has issued the comprehensive report *Alternatives to the Land Disposal of Hazardous Wastes in California.*[25] Based on this report, the Assembly Committee responsible for hazardous wastes held a hearing in February 1982 to explore whether to require the Division of Toxic Substances Control within the California Department of Health Services to issue further controls against landfilling of new wastes.

Thermal devices of various kinds can break down certain organic (hydrocarbon-based) wastes; the rotary kiln incinerator is the most common of these thermal technologies for waste disposal. In contrast to landfills, incinerators can be accommodated in a much smaller area. According to a 1980 report by Clark-McGlennon Associates sponsored by the New England Regional Commission, sites of four to five acres may suffice for an incinerator.[26] Even here, however, ash disposal remains a tough problem, potentially threatening to the environment—and incinerators may pose air emission problems as well. Moreover, alternatives to land disposal of hazardous wastes may be far more expensive than were the landfills and waste lagoons so popular with industry in the past.

Additional hazardous waste disposal alternatives include:

- physical processes, which include sedimentation and flotation (gravity separation), filtration, flocculation (use of chemicals to aggregate smaller particles into larger ones removable by gravity separation or filtration), distillation, and absorption;
- chemical processes, which include neutralization, oxidation, reduction, precipitation, ion exchange, and solidification (use of materials that will render the mixture solid to reduce or prevent leaching); and
- biological processes, which include activated sludge, aerated lagoons, and composting.[27]

As the costs of both production and proper disposal increase, waste reuse or recovery seems certain to expand. EPA has estimated that only about two percent of industrial hazardous wastes are now recovered or recycled.[28] "Waste exchanges" are one potentially useful way to encourage reuse of these materials. Here information can be exchanged so that the wastes of one firm can be transferred to another which can use these wastes as inexpensive raw materials in its own processes.[29] Ultimately, reduction of waste generation through industrial process changes and product substitution should prove the most environmentally sound means of managing hazardous substances, thereby reducing the capacity needed for new hazardous waste management facilities. In the immediate future, however, measures to reduce waste generation are likely to remain so modest that they will not constitute a solution to our hazardous waste problem.

NEW JERSEY'S WASTE SITUATION

The problems of regulating disposal of hazardous waste in New Jersey provide a microcosm of the national crisis. This state ranks as one of the nation's leading waste generators; according to EPA estimates, New Jersey accounts for roughly 8 percent of the nation's total hazardous residues.[30] Because of its small size and dense population, the impacts of waste disposal are quite severe. The dubious distinction of having become "The Garbage State" or "Cancer Alley" stems, in part, from the fact that the state's chemical and pharmaceutical industries are dominant forces in the New Jersey economy.[31] The harmful byproducts of the state's heavily industrialized businesses are an unavoidable fact of life in the Garden State. Given the structure of the state's economy and present consumer preferences, extensive generation of hazardous wastes in New Jersey cannot realistically be avoided.[32]

Accurate data about New Jersey's generation and disposal of hazardous wastes have been quite limited, and only recently have there been determined efforts to improve this information. Ron Buchanan, former chief of hazardous waste management for the New Jersey Department of Environmental Protection (DEP), estimated that about 3.9 million tons per year of hazardous wastes were produced in New Jersey in 1979 (at an estimated 15,500 sites).[33] Of this amount, roughly 3.2 million tons a year were being dumped into the Atlantic Ocean.[34] Governor Brendan Byrne's Hazardous Waste Advisory Commission noted that, of the 720,000 tons of waste generated in New Jersey in 1979 (exclusive of that dumped in the ocean), manifest records showed that generators shipped at least 400,000 tons of waste off-site. The remaining 320,000 tons of waste were disposed of on-site by the waste generators themselves. In 1981, New Jersey had nineteen licensed commercial off-site hazardous waste management facilities. Only one of these provides a broad range of treatment processes: the Rollins Environmental Services facility in Logan Township.[35] The biggest and most harmful area of ignorance concerns the amount of waste that is discarded illicitly. The Governor's Commission admitted, "It is almost impossible to estimate the quantity of waste that is being illegally dumped."[36] One alarming speculation, however, is that as much as 50 percent of the state's waste is being criminally managed, nearly all of it unaccounted for by the manifest system.[37]

Manifest records purport to show that the state is a net exporter of waste (95,000 documented tons per year shipped in versus 113,000 tons per year going out). Obviously, these documents only include those wastes formally identified as such—wastes mismanaged or disposed of illegally would not be included. Indeed, manifest figures on exports and imports of waste are almost certainly greatly understated, perhaps by as much as two-thirds.[38] To the extent that surrounding states strengthen their own supervision of hazardous waste disposal, New Jersey's waste is less likely to be taken out of state, thereby increasing the amount to be disposed of by in-state facilities.[39]

The history of improper disposal of chemical wastes has been profoundly felt in New Jersey. By January 1980, the New Jersey DEP had recorded 247 sites where wastes were known or alleged to have been dumped.[40] The U.S. Environmental Protection Agency has investigated approximately 300 locations in the state for possible groundwater contamination by toxic materials; at least ten of these sites have already been found to present "imminent and substantial" threats to public health.[41] It is likely that the number of these most dangerous sites is significantly higher—by one estimate, between forty and fifty.[42]

Among the many hazardous waste repositories in the state are some that, despite their lesser fame, rival the horrors of Love Canal. The Kin-Buc landfill in Edison, for example, was the largest receptacle for chemical wastes on the East Coast until its closure in 1976. Early that year, Kin-Buc was taking in almost 500,000 gallons of chemicals daily (enough to fill ten Olympic-size swimming pools). In all, as much as 750 million gallons of liquid waste were dumped there. As the last commercial landfill for hazardous waste in the state, Kin-Buc was closed after chemicals were found leaching into the adjacent Raritan River; certain dangerous compounds, like benzene, were also vaporizing into the atmosphere.[43]

The closure of New Jersey's leading commercial waste landfill led to an increased flow of toxics to a facility in Elizabeth, the Chemical Control Corporation, which was licensed to detoxify wastes by incineration. Despite the state's approval of this facility, however, wastes brought there after 1976 began to accumulate. Incineration slowed or even ceased entirely, and the operator simply began stacking the barrels outside, four or five drums high. Before the state placed Chemical Control in receivership in 1979, and began to clean out the most dangerous items, over 50,000 drums had piled up on this three-acre site. National attention was focused on the Elizabeth facility when a major fire engulfed the area in April 1980. If DEP had not already removed thousands of barrels of the most lethal substances from Chemical Control, and if weather conditions had been less favorable, there is little question that a major disaster might well have ensued in the New York metropolitan area.[44] Cleanup of the thousands of drums and the many pools of chemicals left after the fire further strained DEP's resources. The site's surface was finally cleaned by late 1980. The cost of the Chemical Control cleanup exceeded $25 million, covered in large measure under New Jersey's Spill Compensation and Control Act.[45] DEP's management of this and other waste facility cleanup operations has been sharply criticized.[46]

Problems of improper or illegal disposal of wastes continue to demand resources to mitigate these disasters. In response to the nationwide abundance of dangerous toxic sites, Congress enacted Superfund legislation in December 1980. This law provides a $1.6 billion fund for abandoned dump cleanups. In late 1981 EPA designated 115 sites across the country for priority cleanup. Unfortunately, the Fred C. Hart study for EPA estimated that a complete cleanup of "orphaned" sites above would require $22.1 billion, far in excess of original Superfund allocations. The tab for cleaning up all existing chemical dumps posing an environmental threat would be about $44.2 billion.[47]

Just as in the country as a whole, New Jersey's efforts cannot be confined solely to cleaning up past mistakes. A 1980 report by Booz, Allen & Hamilton on New Jersey's waste management system pinpointed six factors that were likely to increase the state's need for additional hazardous waste treatment and disposal capacity:

- existing landfills will reach capacity;[b]
- new RCRA regulations will close some facilities;
- EPA regulations may expand the class of wastes considered hazardous;
- greater enforcement efforts will increase demand for legitimate disposal;
- economic growth will likely increase absolute quantities of waste generated;
- ban on ocean disposal of sludge will go into effect as provided for in federal legislation.[48]

The rules on ocean disposal of wastes, by themselves, are estimated to increase annual demand for commercial hazardous waste treatment and disposal in New Jersey by perhaps as much as 740,000 tons per year, more than doubling the present level of documented waste disposal.[49]

[b]New Jersey as of 1982 does not use landfills for hazardous waste disposal. Some wastes from the state are transported to out-of-state landfills.

THE SITING IMPERATIVE

It is clear that environmentally viable means must be found to dispose safely of these enormous quantities of hazardous wastes. Prior to implementation of RCRA, EPA estimated that only 10 percent of the country's wastes were being handled in "environmentally sound" ways.[50] Rigorous enforcement of RCRA could change this situation, though budget and personnel reductions at EPA make this result unlikely.

In order to implement "safe" handling methods it is also imperative to find sites where the best available hazardous waste treatment techniques can be employed. Former EPA Deputy Administrator Barbara Blum has argued that,

> . . . the simple truth of the matter is that the waste has to go somewhere—we cannot shoot 40 million tons of it off into space. If we do not establish environmentally sound disposal sites, the inevitable consequence is that the waste will wind up in our backyards anyway—but without the controls that would keep it from doing us harm.[51]

Estimates by EPA and the National Solid Waste Management Association suggest that between fifty and sixty new sites for waste facilities will be needed over the next several years.[52]

However, a 1980 study conducted for EPA concluded that sufficient capacity for 1981 did exist in most parts of the country, particularly in the Northeast. Indeed, the report estimated that in 1981 off-site annual capacity of 18.4 million wet metric tons (WMT) would be more than double the demand for off-site disposal.[53]

These findings are borne out by the situation in New Jersey. Here the facilities of the two largest waste disposal firms, Rollins Environmental Services and SCA Services, have been vastly underutilized due to some extent to their high costs. Prior to implementation of RCRA regulations, SCA's hazardous waste recycling facility in Newark operated at a mere 8 percent of capacity. Prior to its temporary closure by DEP for environmental reasons in 1981, Rollins offered the largest range of treatment and disposal processes (including incineration) in the state at its Logan Township plant. Even so, it had done business at only about 50 percent of its capacity.[54] More recently, DEP has placed limits on Rollins' capacity due to periodic mechanical problems and to frequent violations of environmental standards at this facility.[55]

Despite this present underutilization of disposal capacity, there appears to be widespread consensus that this situation will be reversed—nationally, and locally—as the demand for legitimate disposal accelerates in the 1980's.[56] In addition to the six factors noted by Booz, Allen & Hamilton as likely to lead to a marked rise in demand for hazardous waste management facilities, other wastes requiring safe disposal—easily overlooked when focusing on industrial generation—will come from cleanup operations at abandoned hazardous waste sites.[57] These programs will be expanded nationwide as federal Superfund monies are dispensed.[58] As a result, more new sites will be needed—but will they be accepted by residents of the local communities where they are proposed?

The remainder of this study explores various aspects of this siting imperative. Developers of hazardous waste management facilities, local opponents, and state regulatory agencies all have a stake in the outcome of this continuing confrontation. The present situation encourages a fundamental clash of values every time a new waste facility is proposed. Developers justify their proposals with their own calculations of free market economics, and defend their facilities' safety by citing their conformance with applicable federal or state environmental standards. They argue that the waste facility will be of net benefit to society as a whole, and they assume that their property development rights will be upheld by the courts unless opponents can prove conclusively that the facility would violate explicit environmental standards. As already noted, such expectations collide head-on with the wishes and fears of local residents, who stand firm against the proposed waste facility. State regulatory agencies are in a quandry. Unfortunately, they frequently exacerbate rather than resolve the siting dilemma. Mistrusted by many in the local community, and under great pressure to find sites for at least some new facilities, their approval of a facility's technical characteristics and suitability for a particular location carries little credibility with local opponents.

The only feasible resolution to this dilemma appears to reside in creation of a balanced siting process which can somehow accommodate these conflicting interests. Though local residents may never fully accept the presence of a hazardous waste facility in their community, it is vital that they come to respect the *process* by which the siting decision was made. The components of such a decisionmaking process are set forth in some detail later in this book. Allocation of political authority and an effective balance of political power are emphasized in Chapter 4. Negotiated compensation mechanisms designed to redress the inherent imbalance of costs and benefits are the focus of Chapter 5.

Creating a balanced siting process is a high priority on the states' hazardous waste management agenda, and will normally require passage of new state siting legislation such as New Jersey's Senate Bill 1300, approved in September 1981. In essence, an adequate political process for siting hazardous waste facilities is one that attends to the fears of the minority while allowing the majority to obtain new sites. In the absence of such a process, public alienation and anger will grow, and the harmfulness of inadequate hazardous waste management will become ever more significant.

NOTES

1. Douglas Costle and Eckhardt Beck, "Attack on Hazardous Waste: Turning Back the Toxic Tide," *Capital University Law Review* 9 (1980): 426.
2. Joe Cantlupe, "Middlesex Doesn't Want Waste Sites," *New Brunswick Home News* (July 16, 1980). Middlesex County Freeholder Stephen Capestro was commenting on public response to possible siting of a hazardous waste facility.
3. "New Pollution Challenge: A Deluge of Dangerous Chemicals," *U.S. News & World Report* (19 December 1977): 32.
4. Deborah Sheiman, "A Hazardous Waste Primer" (Washington, D.C.: League of Women Voters Education Fund, 1980), pp. 2–3. Also see Grace Singer, *Nor any Drop to Drink: Public Policies Toward Chemical Contamination of Drinking Water* (Princeton, N.J.: Princeton University, Center for Energy and Environmental Studies, 1982).
5. This is a basic theme in the analysis of facility siting problems presented in Michael O'Hare, Debra Sanderson, and Lawrence Bacow, *Facility Siting* (New York: Van Nostrand-Reinhold, forthcoming); draft manuscript, Chapter 1, pp. 6–7.
6. Dennis Ducsik, *Electricity Planning and the Environment* (Cambridge, Mass.: Ballinger, forthcoming).
7. Michael O'Hare, "'Not on *My* Block You Don't': Facility Siting and the Strategic Importance of Compensation," *Public Policy* 25 (Fall 1977): 407–458.
8. Council on Environmental Quality et al., *Public Opinion on Environmental Issues,* 0–329–221/6586 (Washington, D.C.: U.S. Government Printing Office, 1980), pp. 30–32.

9. O'Hare, Sanderson, and Bacow, *Facility Siting,* Chapter 1, p. 3.
10. U.S. Environmental Protection Agency, Office of Water and Waste Management, *Siting of Hazardous Waste Management Facilities and Public Opposition,* SW-809 (Washington, D.C.: U.S. EPA, November 1979), p. iii.
11. Council on Environmental Quality et al., *Environmental Trends* (Washington, D.C.: U.S. Government Printing Office, 1981), p. 85. This report cites information originally presented in: E.C. Lazar et al., "The Potential for National Health and Environmental Damages From Industrial Residue Disposal," Proceedings of the National Conference on Disposal of Residues on Land, sponsored by EPA's Office of Solid Waste (Washington, D.C.: September 1976), p. 196.
12. U.S. Environmental Protection Agency, *Hazardous Waste Generation and Commercial Hazardous Waste Capacity—An Assessment,* SW-894 (Washington, D.C.: U.S. EPA, November 1980), p. III-6. The comparative figure for 1980 was 41.2 million metric tons. To illustrate the potential inaccuracy of knowledge about hazardous waste generation, the 1981 estimates ranged from 27.8 to 53.9 million metric tons.
13. Council on Environmental Quality, *Environmental Trends,* p. 85.
14. U.S. EPA, *Hazardous Waste Information* (Washington, D.C.: U.S. EPA, 1980), p. 2.
15. Council on Environmental Quality, *Environmental Trends,* p. 85.
16. Keystone Center, *Siting Non-Radioactive Hazardous Waste Management Facilities: An Overview,* Final Report of the First Keystone Workshop on Managing Non-Radioactive Hazardous Wastes (Keystone, Colo.: Keystone Center, September 1980), p. 1.
17. Michael Brown, *Laying Waste: The Poisoning of America by Toxic Chemicals* (New York: Pantheon, 1980).
18. Keystone Center, *Siting Non-Radioactive Hazardous Waste Management Facilities,* pp. 1-2.
19. Ed Magnuson, "The Poisoning of America," *Time* (22 September 1980): 63.
20. The RCRA manifest system is frequently described, though erroneously, as one which tracks all hazardous wastes from their point of generation to their point of ultimate disposal (a "cradle-to-grave" tracking system). In fact, under EPA's RCRA requirements copies of the manifests only have to be sent between the shipper and the receiver, not to any government agency. Moreover, it is up to the generator to identify particular wastes as "hazardous" (following EPA's definitions). Some observers fear that the manifest system being established nationally under RCRA may not fully assure people that these wastes are being handled in a safe manner. A few states, in-

cluding several in New England, require that copies of the manifest also go to the appropriate state environmental or public health agency; if the wastes cross state lines, agencies in both states are to be notified. Even here, however, the wastes are not really being monitored "from cradle to grave," an extraordinary complex task involving identification of all wastes and essentially tracking each carbon atom as it moves through the chemical process. Michael O'Hare was very helpful in pointing out the misleading nature of the "cradle to grave" phrase now in common use; personal communication to David Morell, 19 January 1982.

21. EPA estimates that the number of generators involved in the manifest system will be about 275,000; the number of permits for treatment, storage, and disposal facilities will be roughly 30,000. Charles Pierce, "A Look Ahead: Interview with Steffen W. Plehn, Deputy Assistant Administrator for Solid Waste," *EPA Journal* 5 (February 1979): 11.
22. The New Jersey regulations will cover generators which produce less than the federal cutoff of 3,000 pounds per month. Patrick McDonnell, "Go-ahead Nears for Toxin Plan," *Newark Star-Ledger* (2 January 1981): 1, 12.
23. *Land Use Planning Report* (19 October 1981): 320.
24. Jonathan Steeler, *A Legislator's Guide to Hazardous Waste Management,* prepared for National Conference of State Legislators (Denver, Colo.: National Conference of State Legislatures, 15 October 1980), p. 8. Also see Peter Montague, *Four Secure Landfills in New Jersey—A Study of the State of the Art in Shallow Burial Waste Disposal Technology* (Princeton, N.J.: Princeton University, Department of Chemical Engineering and Center for Energy and Environmental Studies, draft 1982), pp. 2–5.
25. State of California, Office of Appropriate Technology, *Alternatives to the Land Disposal of Hazardous Wastes: An Assessment for California* (Sacramento: OAT, 1981).
26. Clark-McGlennon Associates, *An Introduction to Facilities for Hazardous Waste Management,* Prepared for the New England Regional Commission (Boston, Mass.: Clark-McGlennon Associates, November 1980), p. 39.
27. Steeler, *A Legislator's Guide to Hazardous Waste Management,* p. 8.
28. Council on Environmental Quality, *Environmental Trends,* p. 85.
29. Steeler, *A Legislator's Guide to Hazardous Waste Management,* pp. 6–7.
30. U.S. EPA, *Everybody's Problem: Hazardous Waste* (Washington, D.C.: U.S.EPA, 1980), p. 14.
31. According to George Otis of the New Jersey Chemical Industry Council, the chemical industry in New Jersey provides over 130,000

jobs and a $2 billion annual payroll. George Otis, Presentation at League of Women Voters of New Jersey Conference on "The Future of Hazardous Wastes in New Jersey" (Woodbridge, New Jersey, 19 November 1980).

32. Diane Graves, "Hazardous Waste Bill Emerges from the Rubble," *Jersey Sierran* 8 (September-October 1980): 3.
33. McDonnell, "Go-ahead Nears for Toxin Plan," p. 1.
34. Gene Dallaire, "Toxics in the N.J. Environment: Microcosm of U.S. Ills," *Civil Engineering* (September 1979): 82. Governor Brendan Byrne's Hazardous Waste Advisory Commission concluded that ocean disposal constituted a "special case which should not be lumped in with the land-based disposal picture." These quantities are acidic or alkaline solutions that are not highly toxic. New Jersey Hazardous Waste Advisory Commission, *Report of the Hazardous Waste Advisory Commission to Governor Brendan Byrne* (Trenton: State of New Jersey, 1980), p. 16.
35. Hazardous Waste Advisory Commission, ibid.
36. Ibid., pp. 12–13.
37. Paul Arbesman, Department of Environmental Protection, Presentation at Seminar on Toxic Wastes, Woodrow Wilson School, Princeton University (Princeton, N.J.: November 25, 1980).
38. Hazardous Waste Advisory Commission, *Report,* p. 13.
39. Political pressure by other states may also play a role in state efforts to reduce waste exporting. As Congressman Marks of Pennsylvania told DEP officials when they testified before Congress that much of the state's waste was shipped to Pennsylvania, "May I suggest to you that you had better be prepared to come up with a program not to do that any more?" U.S. Congress, House Committee on Interstate and Foreign Commerce, *Hazardous Waste Disposal, Hearings before the Subcommittee on Oversight and Investigations,* 96th Congress, 1st session, 1979, p. 344.
40. Hazardous Waste Advisory Commission, *Report,* p. 65.
41. Robert Rudolph, "EPA Finds 300 Sites Seeping Toxins into State Water Sources," *Newark Star-Ledger* (21 December 1980): 1.
42. Graves, "Hazardous Waste Bill Emerges from the Rubble," p. 3.
43. Hazardous Waste Advisory Commission, *Report,* p. 18; Shawn Tully, "The King of Toxic Waste," *New Jersey Monthly* (November 1979): 102; Rudolph, "EPA Finds 300 Sites," p. 9; Dallaire, "Toxics in the N.J. Environment," p. 83.
44. Magnuson, "The Poisoning of America," p. 64; Tully, "The King of Toxic Waste," p. 105; Jack Anderson, "A Crime Family Surfaces at Chemical Control Dump," *Newark Star-Ledger* (30 December 1980): 14.

45. New Jersey Department of Environmental Protection, *Abandoned Site Cleanup Status Report* (20 November 1980), p. 7. Cleanup activities at Chemical Control were funded largely by the state's Spill Compensation and Control Act.
46. Herb Jaffe, "How Costs Soared in Toxic Waste Cleanup: Shoddy Records Used As Basis to Pay Out Millions," *Newark Star-Ledger* (10 January 1982): 1. This was the first of a series of 12 articles in the *Newark Star-Ledger* on this topic.
47. Dick Kirschten, "The New War on Pollution is Over Land," *National Journal* 11 (14 April 1979): 604.
48. Booz, Allen & Hamilton, *Hazardous Waste Management Capacity Development in the State of New Jersey,* Prepared for the state of New Jersey and the Delaware River Basin Commission (Bethesda, Md.: 15 April 1980), pp. 1–2.
49. Hazardous Waste Advisory Commission, *Report,* p. 13.
50. U.S. EPA, *Everybody's Problem: Hazardous Waste,* p. 15.
51. Chemical Manufacturers Association, "Blum Warns Against 'Chemical Anxiety,' " *Chemecology* (August 1980): 11.
52. U.S. EPA, *Hazardous Waste Generation,* p. viii.
54. Dallaire, "Toxics in the N.J. Environment," p. 86.
55. Graves, personal communication to David Morell, 10 February 1982.
56. Graves, interview, 19 March 1981; George Kush, SCA Services, Presentation at League of Women Voters of New Jersey Conference on "The Future of Hazardous Wastes in New Jersey" (Woodbridge, N.J., 19 November 1980).
57. Arbesman, Seminar, 25 November 1980.
58. The wastes from New Jersey cleanup operations have generally been taken to landfills in Niagara Falls, New York, or to Alabama. David Henderson, Department of Environmental Protection, Presentation at Seminar on Toxic Waste, Woodrow Wilson School, Princeton University (Princeton, N.J., 2 December 1980).

2 RISK, FEAR, AND LOCAL OPPOSITION: "NOT IN *MY* BACKYARD"

Siting of new hazardous waste management facilities is the responsibility of the states and their component local governments. In many respects, this is the most difficult aspect of hazardous waste management. The problem, quite simply, is that no one wants to live near these facilities; as mentioned earlier, the prevailing sentiment is "not in my backyard." This dilemma was summarized in a publication entitled "Hazardous Waste Facility Siting: A Critical Problem" which the federal Environmental Protection Agency (EPA) sent to each state governor in 1980, as a complement to EPA's draft regulations under the Resource Conservation and Recovery Act (RCRA):

> Implementation of the hazardous waste program is not expected to be an easy task. A critical step will be the creation of new facilities employing the most advanced waste management technologies. But to establish newer, improved facilities, sites on which these facilities can operate must be found. Establishing these sites will be an exceptionally difficult task. For while everyone wants hazardous waste managed safely, hardly anyone wishes it managed near them. Yet if the program is to work—if the public health and the environment are to be protected—the necessary sites must be made available.[1]

In the past, including the recent past, most hazardous waste facilities were in fact "dumps," both traditional open landfills and the more modern covered "sanitary landfills" or lined "secure landfills" with engineered leachate collection systems. Since most of the public's experience has been with these dumps, which have often leaked their dangerous contents into the surrounding environment, opposition to new waste facilities has intensified greatly in recent years. Few make a clear distinction between these traditional approaches to disposing of toxic wastes on the land and the new generation of high-technology hazardous waste facilities described in Chapter 1. While these, too, may at times warrant local opposition, lack of clarity over the type of waste facility contributes markedly to today's siting problem.

COMMUNITIES IN FEAR

Proposed new hazardous waste facilities typically encounter intense public opposition. At the heart of such opposition lies fear, fear of being the victim of another environmental disaster. Publicity about Love Canal, Chemical Control Corporation, and other serious episodes in various locations around the country has fed these fears, which are—by most available estimates—the dominant reason why so many citizens decide to stand against a new hazardous waste management facility proposed for their community. In the words of a 1979 study by Centaur Associates for EPA, "Opposition is rooted in the fears of major and long-term risks posed by facilities to the health and welfare of the surrounding community."[2]

Some opposition to hazardous waste facilities emerges from factors other than fear, however, and it is important for policy-makers to distinguish between opposition based on fear and opposition using this rhetoric but really resting on other causes: the facility's impact on property values, the community stigma of becoming a "regional dumpsite"; heavy truck traffic; noise; the visual impact of a waste facility in the neighborhood; and so on.[3] For example, some people in the proposed locale may become convinced that improved technology and monitoring and control at a modern waste facility really will assure their safety. At the same time, they might worry that the person to whom they might eventually want to sell their house in the future would not be so convinced—but would instead be afraid of the facility's dangers. Those who oppose a proposed hazardous waste facility for such legitimate reasons, however, may well realize that they can mobilize much more support for their cause from others by raising the specter of fear than by simply saying "I want to live in a nice community, not near a dump." On the whole, of course, many people are *truly* afraid of these facilities—and with good cause. Yet others who express such fears may have different agendas. This distinction is important because improvements to a facility to reduce its risk may still leave significant community opposition based on its other impacts, and because monetary compensation mechanisms may be particularly appropriate in dealing with opposition based on such factors.

Moreover, as just noted, some of the opposition to new hazardous waste management facilities is due to people's confusion between the old generation of tremendously risky chemical dumps, of which thousands still remain, and the new types of well-monitored treatment and disposal facilities. One can certainly understand people not wanting a new chemical dump in their backyard. But many people oppose a new waste incinerator or a new waste trans-shipment terminal simply because they fail to distinguish these modern facilities from their poorly managed predecessors. Many of these same people might not oppose the siting of a new petro-chemical facility at the same location, even though this complex technological entity might be very similar to a modern waste management facility. In this sense, image and semantics are terribly important—the very phrase "hazardous waste" connotes problems to a concerned citizenry. Indeed, successful development of state and national siting strategies rests on the ability of government and industry to make these distinctions between old and new facilities clear to the public.

Furthermore, both government and industry lack credibility—a fact fully understandable given their past actions. Many people lack trust in the ability of government—federal, state, or local—to minimize the risks from hazardous waste facilities by enforcing adequate environmental and health standards for such potentially dangerous facilities. Industry's continual assurances that risks will be kept low are likewise discounted due to past instances of industry's carelessness and irresponsibility.[4] Often public mistrust and suspicion of developers increase during a siting conflict, as inconsistencies become evident between the developer's words and his actions and as he proves reluctant to provide the local populace with accurate, timely information about the project.[5] The credibility gap facing both government and industry will have to be closed before successful siting strategies can ever hope to evolve.

As a result of this convergence of factors, most attempts to site hazardous waste facilities within the past few years have met with vehement opposition in many states. In one New Jersey municipality, East Windsor Township, for example, simply the *possibility* of becoming the site of what was assumed to be a chemical dump (following the issuance of draft site selection criteria in July 1980 by the Delaware River Basin Commission and the New Jersey Department of Environmental Protection) prompted residents of the town to turn out en masse to a meeting in a school gymnasium to obtain information from local officials and to express their opposition. At least 1,000 residents attended this meeting, making it the largest ever held in the township.[6]

One such siting scare also took place in New Jersey when residents of Alloway Township discovered in December 1980 that Envirosafe Services, a subsidiary of IU (International Utility) Conversion Systems, Inc., planned to build a 400-acre landfill on land acquired by a "front" developer ("Fox Hunt Farms") who many had thought planned to build housing in the area. After having invested $500,000 in the project, the waste management firm abandoned its plan in the face of rising public opposition. Although Envirosafe officials expected that they would have to persuade the township to accept the dump, they discarded their plan when they realized the extent of local opposition and the probability of lengthy litigation.[7]

The experience of 640 residents of Jackson Township further illustrates the fear that New Jersey residents have of living near hazardous wastes. In the Legler section of this community, a municipal landfill was sited in 1971 despite residents' protests. In 1979, the state established that illegal disposal of hazardous wastes at this site had caused 146 residential drinking water wells to become contaminated by harmful organic chemical compounds.[8] The residents who were affected in Jackson, not surprisingly, have voiced considerable anger over state and local governments' role in allowing the landfill to continue to threaten their health. The municipality and the state have encountered administrative problems in obtaining and financing alternative uncontaminated water supplies.[9]

Widespread opposition to siting of new hazardous waste facilities has not been confined to New Jersey, of course. A 1979 report prepared for EPA—"Siting of Hazardous Waste Management Facilities and Public Opposition"—describes nine cases across the country where public opposition either halted the siting of a facility or stopped it from continuing to operate.[10] Citizens in the town of Hooksett, New Hampshire, for example, have resisted the siting of a waste solidification facility by the Stablex Corporation. In February 1981, the Hooksett Planning Board voted to reject the Stablex application as contrary to local zoning ordinances. Stablex subsequently appealed this local veto to the Merrimack County Superior Court.[11]

Elsewhere in New England, residents of Westford, Massachusetts, successfully resisted an Industrial Tank (IT) Corporation plan for a hazardous waste treatment plant that would have employed technology new to the United States. This facility was designed to incinerate, neutralize, and detoxify as much as 500,000 tons of wastes per year. Despite new state siting legislation designed to facilitate successful siting, residents in this community of 14,000, following an emotional town meeting attended by 3,000 citizens, eventually were able to persuade the owner of a local granite quarry not to sell his land to the waste disposal company.[12]

> The Westford opposition was reportedly based on two objections to IT's proposal: (1) residents felt the siting process had moved so quickly that it was "difficult for amateurs to keep up" with the technical issues involved; and (2) they felt it unfair that one small community should have to host a large facility that would process wastes originating from the entire New England region.[13]

In New York State, a 2,800-acre spot in the town of Sterling in Cayuga County has been proposed by the state as the location for a large new hazardous waste treatment plant that would dispose of much of the state's wastes. This site, chosen by use of state siting criteria, emerged from a list of thirty-one finalists due to its favorable location on Lake Ontario (the water from which would be used to dilute the plant's effluents). However, many of Sterling's 3,500 residents have opposed the idea of becoming "New York's chemical dump"; indeed, a previous nuclear power plant proposal for this same site, while defeated, had enjoyed significantly more support (due primarily to the higher tax benefits envisioned by the town). As emotions built over the proposed hazardous waste facility, one member of the local citizens group, Citizens Concerned About Sterling, stated, "People are talking about sabotage . . . We're not talking about any radical element. This is a very conservative town, but people are very upset."[14] In late 1981, New York abandoned this siting proposal. Governor Hugh Carey announced that he would appoint a special panel to explore "the complex legal and technical problems associated with hazardous waste so that necessary facility siting and regulatory programs can receive public support." As Department of Environmental Conservation Commissioner Robert Flacke put it: "We had to admit defeat in the face of strident public opposition." In Carey's words: "It's the old story. Nobody wants it in their backyard, and it winds up in nobody's backyard."[15]

Opposition to establishment of hazardous waste facilities has not been limited to the Northeast either. Citizens in Starr County, Texas, fought a landfill proposal that would have been sited at one of the most geologically suitable spots in the state, perhaps in the entire country. The principal basis for opposition in this case was Starr County's refusal to become the "dumping ground" for wastes generated in the Houston-Galveston petrochemical complex.[16] In Wilsonville, Illinois, an Earthline, Inc., landfill which had been open for five months with minimal opposition became the proposed disposal site for some soil from Missouri contaminated with polychlorinated biphynels (PCB). Intense local opposition quickly surfaced, and local reaction to the landfill shifted rapidly away from the formerly neutral attitude about the facility. Violence was threatened, and legal proceedings by the town resulted in the closure of this site.[17] In Los Angeles County, California, the extension of an existing landfill operated by the county brought forth objections from local residents as early as 1973. This opposition came from residents of a nearby residential development; their objection rested on aesthetic concerns: "the effect the new landfill would have on their view of the hills."[18] In Puerto Rico, a public interest group—Puerto Rico Industrial Mission—has used legal tactics to resist siting of a landfill in the community of Ponce. The group has objected to the appropriateness of landfill operations on the densely populated island and has questioned the industry plan's technical soundness. Citizens in Puerto Rico were also concerned about their area becoming a dumping ground for the continental U.S.[19]

Perhaps the most striking indication of the powerful public opposition to the risks of living near hazardous wastes was the response of a community leader from Warrenton County, North Carolina, to a proposed facility to dispose of 40,000 cubic yards of soil contaminated with illegally dumped PCB's. In order to keep out these wastes, he implored citizens to "stand up in front of the dump trucks and bulldozers and even give up our lives."[20] Those who hope to site hazardous waste facilities—both in industry and in government—must come to terms with this type of formidable obstacle.

DIVERSITY OF ISSUES: BRIDGEPORT AND BORDENTOWN

Two siting cases from New Jersey illustrate several of the key principles in local opposition to siting new hazardous waste management facilities.[21] While the siting process was essentially the same for both proposals, an on-site industrial landfill was approved while an off-site regional landfill was rejected in the face of intense citizen opposition.

Between 1976 and 1978, the Monsanto Chemical Company successfully sited a new hazardous waste landfill at its Delaware River Plant near Bridgeport, New Jersey, in an unincorporated area within Logan Township of Gloucester County (see Figure 2–1). The landfill is located on the plant's grounds and disposes of chemical sludges generated at the plant itself. There was no organized opposition to this siting proposal. In contrast, in 1978 the SCA Corporation's subsidiary, Earthline, attempted to establish a commercial hazardous waste landfill in Bordentown, New Jersey. This location is also on the Delaware River, between Philadelphia and Trenton, about thirty-five miles north of Bridgeport. Here citizen opposition arose quickly, and led local officials and members of the state legislature to oppose the facility. In 1979, Earthline's application was denied by the New Jersey Solid Waste Administration (SWA), the state's permit-issuing authority.

In both of these siting attempts, the institutional procedures governing facility location were the same. For several years, siting of hazardous waste facilities in New Jersey has been exempted from local zoning controls. Instead, such siting has been regulated by the Solid Waste Administration, a component of the state's Department of Environmental Protection (DEP). SWA's dual mandate called for the agency to meet both the wishes of waste facility developers and the environmental quality concerns of a locality. (See Appendix B for an up-to-date summary of siting mechanisms in New Jersey.)

Another factor common to both siting efforts was the possibility of major environmental contamination. Both landfills were designed to contain toxic chemicals, and both were proposed for environmentally sensitive watersheds. The Monsanto facility was sited within 150

Figure 2-1. Bordentown and Bridgeport, New Jersey.

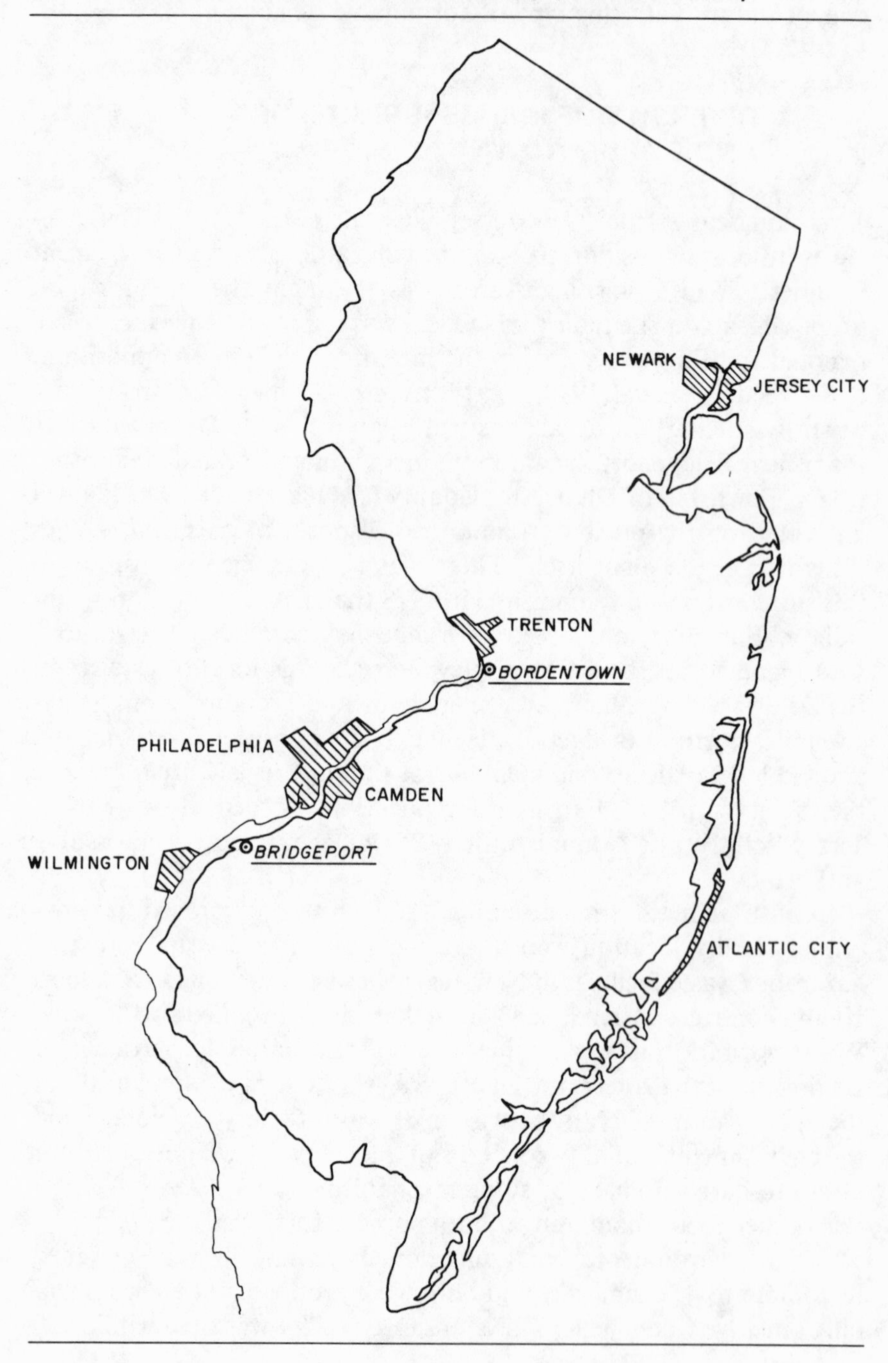

feet of the Delaware River; perhaps even more dangerous, the proposed Earthline landfill was not only near the Delaware River but was located over a major aquifer providing 80 percent of the groundwater pumped in all of Burlington County.[22]

Both Logan Township and Bordentown had been the scenes of heated siting controversies prior to the hazardous waste landfill siting attempts of the late 1970s. Between 1975 and 1978, Dow Chemical Company had attempted—without success—to establish a large chemical storage facility in Bordentown. After years of growing citizen opposition, spearheaded by a local environmental group established to fight the Dow proposal, Bordentown officials passed an ordinance which effectively banned the facility's construction. The twin-reactor Newbold Island Nuclear Generating Station had also been proposed for Bordentown. This nuclear facility was rejected in 1973 during the licensing process following heated public discussion in the newspapers and at Atomic Energy Commission (AEC) hearings. The primary issue raised in this controversy was density of population in the area.[23]

Logan Township is the site of Rollins Environmental Services, a hazardous waste processing facility established in 1970. This facility has engendered controversy ever since it opened. Following a serious explosion at the plant in December 1977, the Rollins facility encountered renewed local opposition.[24] The state environmental agency ordered the Rollins facility closed for some months in 1980 due to another explosion and violations of several statutes. In addition, two major energy facilities were proposed in the 1970s for Logan Township, both to be on the Delaware River near Burlington: a Shell Oil Company refinery, and a terminal to unload and store imported LNG (liquefied natural gas). Both proposals were defeated due to local opposition. Here, too, a nuclear power plant had been proposed: the Burlington Nuclear Generating Station. Inadequate supplies of cooling water and state concern regarding overall population density near this twin-reactor complex led the utility and the AEC to shift the facility to an alternative site on Delaware Bay, where it was approved as the Salem Nuclear Generating Station.[25]

Despite their similarities, the Bridgeport and Bordentown hazardous waste siting cases differ in several important respects, and these differences ultimately explain why the two siting attempts were resolved in opposite ways. In considering the differences, it is clear that although the issue of risk to health and safety was a central aspect of both of these siting processes, public reaction was also grounded in factors more complicated than simply a judgment as to whether each of the proposed facilities was safe enough.

These additional factors, which help to account for the acceptance or rejection of a facility, relate both to the siting *process* itself (for instance, whether it was seen as being "fair" or "unfair") and to the anticipated *results* of the siting attempt (that is, whether the facility, on balance, was seen as harmful or beneficial to the community). Public response may be premised on the "means" or the "ends" of siting, or on both. Naturally, factors of both types are typically involved in any siting attempt; the Bridgeport and Bordentown cases are particularly instructive in illustrating how a diversity of issues can shape the outcome of a siting decision for a controversial new facility.

The Siting Process

One of the crucial determinants of the success or failure of a siting proposal is the manner in which the decisionmaking process is carried out and the resulting form of public reaction. Among the factors which relate to this process are the actions of organized interests and the perceived legitimacy of the site review effort.

Actions of Organized Interests. In Bridgeport and Logan Township, local citizens made essentially no attempt to organize opposition to Monsanto's proposed on-site landfill. The company filed its application in 1977 with SWA to establish a new chemical landfill on the grounds of its manufacturing plant. The six-acre site had earlier been used by the United States Army Corps of Engineers to deposit materials dredged from the Delaware River channel.[26] The landfill was designed to accept thirty to forty truck loads per week, primarily liquid sludge from Monsanto's manufacturing operations.

The state agency notified local agencies and officials, including the mayor of Logan Township, the Logan Township health officer, the Logan Township Environmental Commission, and the Gloucester County Planning Department. Monsanto's subsequent actions seem to have played a vital part in winning local support. First, the company discussed its plans with local officials, including some members of the Logan Township Environmental Commission. These talks evidently convinced them that acceptance of Monsanto's proposal would be in the best interest of the town. As a result, the landfill received the support of these public officials in Bridgeport and Logan. Indeed, after meeting with Monsanto, the Environmental Commission reportedly endorsed the application[a] and requested that SWA expedite the review process.[28]

Monsanto likewise successfully handled public and official scrutiny of its proposal by *negotiating* on the specifics of its proposal in order to increase the acceptability of the facility; these changes were primarily designed to reduce its hazards. The company had originally proposed a landfill with a single clay liner. In response to SWA's concern about the liner's permeability, Monsanto agreed to install a double liner structure with a leak detection system between the two liners. As built, the landfill includes an 18-inch clay primary liner and a 12-inch clay secondary liner. In addition, a layer of special material produced by Monsanto (BIDIM) was placed below the secondary liner to lessen the chance that uneven settling of the subsoil would destroy the integrity of the clay liner.[29] The company's willingness to make changes to lessen risk of waste leakage clearly had the effect of increasing Monsanto's perceived credibility in its design and promotion of an acceptable facility.[b]

The public hearing held by SWA to discuss the permit application—a standard agency procedure—did not elicit from the public-at-large any substantial degree of opposition.[c] Instead, public response seemed to indicate an attitude of qualified acceptance:

> In general, most comments appeared to indicate at least general support for Monsanto and/or the proposed landfill. Most questioners sought additional information about the facility, particularly contingency plans in the case of accidents and post-closure provisions. Monsanto provided responses to all comments either providing additional information about the facility or indicating Monsanto's existing and future commitments to the Bridgeport plants and to Logan Township in general.[31]

The public hearing also made evident the extent to which community officials supported the proposed landfill; their opposition was limited

[a]Some observers report instead that only one member of the Environmental Commission actually expressed support, or that the mayor may have claimed that the commission supported the proposal—whereas in actuality the Environmental Commission as a whole did not support it.[27]

[b]Note, however, that the Earthline landfill proposed for Bordentown contained far more engineering features than did Monsanto's landfill in Bridgeport, yet the Bordentown facility was defeated.

[c]There is some concern, however, over the adequacy of this public hearing as a true gauge of community sentiment on this siting proposal.[30]

to the Gloucester County Environmental health coordinator, whose primary concern was contamination of nearby water.

This public hearing was essentially an "administrative ritual"[32]—SWA approval of the facility seemed already assured. Nevertheless, it seems apparent that those few citizens who opposed the facility would have been hard-pressed to establish effective resistance to its siting. Much of Monsanto's success can be attributed to the company's careful actions during the siting process, as the company lined up support among key local actors and responded to criticisms of the proposal by expressing some willingness to improve it. The company's informal influence over some local officials may have been substantial.

Unlike the Monsanto experience, the SCA (Earthline) proposal to locate a hazardous waste landfill in Bordentown met with vehement opposition by officials and residents alike. Undercurrents of this opposition can be found in events that occurred prior to the application for the hazardous waste landfill. First, SCA was no stranger in Bordentown. Intense conflicts had arisen between the town and another SCA subsidiary, Interstate Waste Removal, which operated the town's 100-acre sanitary landfill for municipal wastes. The EPA study notes,

> Prior to the Earthline permit application, an SCA subsidiary had generated anger and resentment in the Bordentown area. Local leaders perceived SCA as a high-powered, politically powerful firm that warranted little trust.[33]

In addition, the defeat of a siting attempt by the Dow Chemical Company in Bordentown just prior to Earthline's hazardous waste facility proposal was instrumental in establishing a climate of local opposition to facilities seen as environmentally suspect. Thus, when SCA/Earthline entered the scene, the organized machinery for public opposition was still intact.

The Dow case was a classic instance of effective organization and persuasion by a small group of citizen leaders. When Dow first proposed to build a chemical tank farm in Bordentown in 1975, local officials supported the plan enthusiastically. After all, it was a new ratable that would have brought in about $1 million in taxes.

Despite official support for the Dow proposal, resistance arose among a small group of middle-class professionals. These individuals came together in an organization called Help Our Polluted Environment (HOPE), whose principal concern became the preservation of local environmental quality. A HOPE-sponsored telephone poll of

Bordentown residents showed strong opposition to the Dow tank farm. The group later gathered signatures on a petition opposing the facility. After a nonbinding referendum that went two to one against the Dow proposal, Bordentown officials finally responded to citizen opposition and passed an ordinance banning the chemical facility. In order to establish the basis for this opposition, HOPE had managed to gain respect within the community. The group capitalized upon media events, used scientific expertise from within the community, and approached the citizens of Bordentown on a face-to-face basis to demonstrate the problems associated with the planned facility.

The success of HOPE's resistance to the siting of the Dow facility in Bordentown illustrates the importance of *organized* opposition. Such opposition was a critical factor in the rejection of Earthline's hazardous waste proposal. The first action taken by HOPE in response to the Earthline application was to hold an informational meeting in which the risks of the facility were discussed. HOPE used this occasion to call attention to events in other parts of the country that indicated the dangers of living near toxic waste (Love Canal) and to other communities' perceptions of the SCA corporation as a local liability (in Wilsonville, Illinois, and Lewiston, New York). In order to focus local attention on the SCA-Earthline facility just before the state hearing, HOPE organized a more unique opposition tactic, a week of "anti-Earthline, pro-people" events. These activities included a parade and rally which lasted several hours.[34]

By the time the New Jersey Solid Waste Administration held its two-day public hearing in Bordentown, the town was solidly against the proposal:

> Every available source strongly indicates that by the time of the hearing area residents were intent on stopping Earthline. As the *Trenton Times* reported the morning of the first day of the hearing, "suddenly, almost everyone here [that is, in the Bordentown area] has become an environmentalist." The article quoted a fairly broad spectrum of local residents, including student leaders at the high school, as being opposed to the facility, and referred to a petition against the facility signed by almost 80 percent of the high school student body.[35]

The proposed landfill's proximity to Bordentown High School heightened public opposition to it.

This widespread local opposition should not be entirely credited to the work of HOPE, however. From the time of SCA's application,

Bordentown officials had treated this proposal with extreme caution, an attitude born of the Dow controversy. These officials were especially committed to a careful and complete study of the SCA application, a situation which led to conflicts with the state agency over the amount of time local officials would be allowed to comment on the proposal. SWA, in its role as both regulator and promoter of waste facilities, appears to have sought to expedite the siting of this landfill in Bordentown in order to expand the state's hazardous waste disposal capacity. Two extensions of time were granted; but when a third request was denied, the city filed a suit against SWA to suspend its permit review until Bordentown City's water supply consultant had had an opportunity to examine the potential impacts of the hazardous waste landfill on the community's water supply. This suit resulted in a settlement allowing a fairly lengthy postponement of the state agency's decision on the application.

Over the course of the review of the SCA proposal, several studies were made. One of the most important, by the Stevens Institute of Technology, was commissioned by Bordentown Township and Burlington County. The report was highly critical of the proposed landfill, questioning in particular the thoroughness of the SCA plan and citing insufficient attention to prevention of aquifer contamination. This report resulted in a significant opposition move by two state assemblymen from the Bordentown area who petitioned the governor to deny the SWA permit, stating that "the tragedy of Love Canal cannot be permitted to repeat itself in New Jersey."[36]

Legitimacy of the Siting Process. The widespread opposition to the SCA application also resulted from local dissatisfaction with the siting procedure itself. The decision as to whether to allow SCA to begin disposal of hazardous wastes rested at the state level with the Solid Waste Administration. Thus, unlike the Dow Chemical case where the town was able to act on its own to exclude an undesirable facility, local officials and residents did not wield such authority over SCA's application. Local residents were therefore apt to feel more frustrated during this landfill siting attempt due to their lack of direct input into the decisionmaking process. This inability to render a local decision on a project that so vitally affected local interests undoubtedly helped opponents of the facility extend the basis for resistance beyond specific concerns about environmental quality to the broader issue of citizen access to government decisions—an appeal to deeply held beliefs about the importance of local consent in the American political system.

Some Bordentown residents also regarded the siting process as suspect due to problems with the state decisionmaking body, most notably SWA's apparent bias in favor of successful siting and the many limitations that characterized this agency. The EPA report notes a lack of *credibility* on the part of the state government agency:

> Opponents doubted SWA's ability to make an independent, reasoned judgment on the application. This was based largely on assumptions that SWA was predisposed to approve the application (or at least be more concerned about waste disposal than public health), that SWA had no guidelines (particularly siting criteria) by which to judge the application, and that SWA's limited qualifications and resources prevented a thorough, independent review.[37]

Objections to the siting process were premised, then, not simply on the perception of an unfair balance between local and state input into the final decision but also on the belief that the existing state facility-siting policy was inappropriate. Not only did the Solid Waste Agency lack criteria with which to judge an application, but—as HOPE pointed out at the public hearing on the Earthline application—at the time of this siting process there were no state or federal regulations governing hazardous waste management. This situation meant that "industry was determining the nature of hazardous waste management practices."[38] Hence, local sentiment favored the establishment of a unified hazardous waste management plan as a necessary preliminary to specific facility proposals such as the SCA plan for a hazardous waste landfill. Indeed, this argument proved so compelling that after the public hearing, the commissioner of DEP announced that the state would not render a decision on any further chemical waste landfill requests in New Jersey until the department had issued new overall hazardous waste regulations.

Overview of Siting Process Differences. This comparison of the Bridgeport (Logan Township) and Bordentown siting processes makes clear the differences which played a large part in the acceptance of one facility and the opposition to and defeat of the other. While one facility was on-site and the other free-standing, public officials and organized interests assumed dissimilar roles in these two cases. In Bridgeport and Logan, local officials, including members of the township's environmental commission, endorsed the proposal; no organized group surfaced to oppose the siting attempt. In Bordentown, on the other hand, the SCA proposal encountered resistance from all quarters at the local level. Local officials first scrutinized the plan carefully and then decided to oppose it, while a highly influential citizen environmental group already in existence (HOPE) organized widespread opposition among community residents.

The credibility of the parties involved in the decisionmaking process was another significant factor. In Bridgeport, Monsanto had a good deal of success in gaining the trust of the community by providing information and clarification of its application and by being willing to negotiate engineering improvements in its landfill siting proposal. Because its large chemical plant was already operating in Logan Township, the company was well known to local residents and the town's officials. Just the opposite was true in Bordentown: SCA inspired little trust in the community both because of its past actions (disputes over the company's sanitary landfill in Bordentown, and the intensely negative experiences described by Wilsonville, Illinois, representatives) and the actions it took during the actual siting process in Bordentown. In contrast to Monsanto's negotiating approach, SCA adopted a more confrontational mode of response to community comment. The EPA report indicates, for example, that "Earthline's attorney also went on local radio and television shows to 'debate' opponents of the facility. According to opponents these debates did little to change anyone's mind."[39] The credibility of the state decisionmaking agency was also at issue in Bordentown. Local residents felt that their fate was in the hands of individuals who were predisposed to approve the industry's landfill request and who, due to a lack of clear criteria and agency resources, did not appear able to provide satisfactory evaluation of its drawbacks.

Finally, the response of the Bridgeport/Logan and Bordentown communities to the siting procedure itself was quite different, though the rules governing this process were essentially the same. Bridgeport residents did not object to the state-dominated decisionmaking process nor to the "token" input afforded by the SWA public hearing. In Bordentown, however, the legitimacy of this process was called into question due to a lack of local input into the final decision and the absence of a carefully constructed application review by the permitting agency. A more legitimate process, according to the Bordentown critics, would have included explicit criteria for judging a landfill application, criteria that would be utilized only after establishment of an overall regulatory mechanism for management of hazardous waste.

While the opponents of the Earthline proposal complained that their interests would not be satisfactorily taken into account by the state decisionmaking agency, it is ironic that ultimately their objections *were* heeded: SWA rejected the application. Indeed, according to the EPA report, both sides felt until the end that SWA would respond favorably to the developer's plan to increase the state's hazardous waste disposal capacity.[40] It is clear from this example, among others, that although a siting process may technically establish the power to preempt local opposition to a facility plan, the political impact of this local opposition at the state level should not be underestimated.[41]

The Result

It is also essential to look beyond the siting process itself, and beyond the actions of the various actors, to examine the substance and context of the siting decision. Undoubtedly the most prominent concern in siting any hazardous waste facility, as mentioned previously, is the question of risk—in other words, is the facility safe?[d] As the Bridgeport and Bordentown cases both demonstrated, however, several other impacts associated with facility siting plans can be instrumental in shaping public response. This case thus reinforces the point that the response of residents of a community to a facility proposal is predicated on the degree to which the facility is seen, on balance, as either beneficial or harmful. Aside from the issue of risk, a wide array of costs and benefits can determine how the local public will respond to a proposed new hazardous waste facility.

Economic Factors. One obvious indicator of the relative advantages and disadvantages offered by a hazardous waste facility is its

[d]The main risk in the Logan Township case was whether leachate from the landfill would contaminate the nearby Delaware River (150 feet away) or Birch Creek (1,000 feet away). Among the dangers were the closeness of these water bodies, the subsidence potential of the landfill area, and the higher water table at the site.

In Bordentown, the principal risk was contamination of the Magothy-Raritan aquifer, the groundwater source for the whole area. The technical debate thus revolved around the inherent geological suitability of the site and the sufficiency of SCA's proposed liner ("two feet of compacted clay, a 40-mil Hypalon liner, six inches of sand, and a minimum of ten feet of clay beneath the sand").[42]

economic impact on a community. The economic effects of the proposed Monsanto and Earthline facilities on their respective communities would have been quite different. In Logan Township, the new hazardous waste landfill was described by Monsanto as being indispensable to continued operation of its chemical plant, a claim that was expressed clearly by the company throughout the siting process. Because Monsanto employs 180 workers at this plant (many of whom live in the Bridgeport area) and pays taxes to Logan Township, the local community faced potentially serious and obvious economic dislocations if the facility were not approved.

In sharp contrast, the Earthline proposal for Bordentown promised virtually no economic benefits to the community. It was to be a free-standing ("off-site") hazardous waste landfill, would have hired few if any local residents, and would have served industry from all of New Jersey, not just from the Bordentown-Trenton area. While defeat of the SCA siting proposal meant loss of the facility's potential tax payments, this was an opportunity cost somewhat less evident to local residents than a decline in existing tax receipts.

Noneconomic Costs. Moreover, the costs and benefits of a facility should not be confined to purely economic factors.[43] People may oppose a facility because it would be an obvious nuisance, the harm here being in terms of amenity costs rather than economic ones. In some instances of hazardous waste facility siting, for example, potential odors have triggered citizen resistance to a facility.[e] In other cases, local residents have opposed a facility because of the traffic associated with it. The possibility that such nuisances—or other noneconomic costs—would accompany completion of the Monsanto landfill was not a major concern of most Bridgeport residents during the siting process (although subsequent to the landfill's construction, there have been complaints about odors). In large measure issues such as odor and noise and traffic did not enter the siting debate because of the facility's favorable location: "It is relatively isolated, only a few residences nearby, and the facility is not visible from U.S. 130, which passes by the plant."[45] Due to its relatively isolated locale, the facility was expected to have a limited adverse impact on the community.

[e]Lee Oldham, vice president of the Hearthstone Homeowner's Association in California, testified at a 1980 state hearing: ". . . every morning we can smell not garbage, we can smell chemicals in our community. And I don't know what chemicals they are. I'm not a chemist. But my wife feels very distraught when she sends our two little boys to school smelling those chemicals at 7:30 in the morning."[44]

Land-Use Compatibility. The belief that the Monsanto landfill was sufficiently isolated to have limited impacts on the community seems to have been part of a broader feeling by Bridgeport and Logan Township residents that this facility was basically compatible with existing land usage. The burial of hazardous chemicals in the vicinity of Bridgeport was obviously not a new and alarming land use. The new landfill was to be sited on the grounds of the Monsanto plant itself, which already had a waste landfill. (Indeed, the fact that this existing landfill was nearing capacity had prompted the siting of this new landfill.) Moreover, the Monsanto operation was located in an industrialized locale, close to other petrochemical facilities.

Residents of Bordentown, on the other hand, viewed Earthline's hazardous waste landfill in a much different light. Although this facility was to be built on the site of an existing sanitary landfill, it was much less remote than its Monsanto counterpart: it would have been directly across the street from Bordentown High School and near the New Jersey Turnpike. In addition, although the new activity would not have worsened the sanitary landfill's visual characteristics, continuation of this landfill's regular (that is, nonhazardous) waste disposal operations was important to the Bordentown area. In its decision to reject the partial conversion of the landfill into a hazardous waste receptacle, the SWA finally concluded: "The secure landfill would reduce needed space already committed to use as a sanitary landfill. The proposed facility would be inconsistent with and disruptive to the area's solid waste management planning."[46]

The New Jersey Turnpike Authority also opposed the facility's siting as incompatible with existing land use. The agency argued that smoke from fires at the site would result in dangerously reduced visibility on the road. Proximity to the turnpike was thus both an asset and a liability for the Earthline plan: an asset because it allowed easy access to the site, but a liability because of potential impediments to other traffic.

Overall, the issue of appropriate land usage was a prominent factor in the SCA case. Ironically, SCA had selected the Bordentown location partly because of its assumption that the presence of an existing landfill at the proposed site could be taken as a sign of public acceptance of intensified waste disposal operations, thereby limiting the potential for public opposition. Nevertheless, although the community had reluctantly assumed the task of waste management,[47] it was *not* prepared to make the transition to *hazardous* waste management. Much of the reason for this resistance stemmed from concern over whether a hazardous waste landfill at the proposed location would be a suitable use of the community's land; considerations such as the presence of the nearby high school led many to believe that this usage would prove quite unsuitable.

Community Perceptions of Costs and Benefits. Given these differences offered by the Monsanto (on-site) and SCA (off-site) facilities, it is not surprising that the former received site approval while the latter did not. It is also important, however, to consider the framework with which each community *perceived* these costs and benefits. The objective costs and benefits of a proposal are less important politically than individuals' perceptions of these costs and benefits. Even if the impacts associated with the Bridgeport and Bordentown cases had proved to be identical, it is clear for the reasons stated earlier that Bridgeport residents would have been more likely than those in Bordentown to consider these impacts acceptable.

An essential aspect of the Bridgeport case is the fact that Monsanto's corporate presence in the area was seen as a significant benefit for the community. Clearly, the dominant political values in the area paralleled the interests of Monsanto. Bridgeport's population is geographically dispersed and consists primarily of working-class families. "Major employers in the area are chemical and petroleum companies with a number of facilities along the Delaware River," according to the EPA study.[48] Citizens in Bridgeport were thus likely to be sympathetic to Monsanto's requests for a new landfill even if they had no direct economic interest in the plant:

> The consensus among those interviewed was that Monsanto's stature in the community obviated the development of opposition. Monsanto has been in Bridgeport almost 20 years, provides a substantial number of jobs, has the confidence of community leaders, and was responsive to community concerns.[49]

An important consequence of this identification by the Bridgeport community with Monsanto and with the basic interests of the chemical industry was that local residents were not overly sensitive to the disadvantages of living near a hazardous waste landfill. Although many did not necessarily embrace the landfill itself, most concerned local residents were mindful of the overall net benefits to the community of accepting Monsanto's siting plan.

Again, the Bordentown situation was quite different in this respect. Whereas Bridgeport's economy is based upon the petrochemical industries, "the employment base of the Bordentown area is broadly dispersed among economic sectors within the greater Trenton metropolitan area."[50] As a result, in no sense were the interests of the citizens of Bordentown allied with those of SCA. Quite the opposite was true since, as indicated earlier, the SCA firm lacked credibility. Thus, residents and local officials were prone to be skeptical of company assertions, to pose tough questions, and to perceive the net drawbacks of living near a hazardous waste landfill. With little reason to see things from SCA's perspective, Bordentown residents considered the cost of this facility to far exceed its benefits.

Equity. When residents of Bridgeport and Bordentown assessed the result of these two siting proposals, they also seem to have arrived at very different conclusions about the *fairness* of being asked to live near these new hazardous waste landfills. Monsanto was fortunate in this regard insofar as its proposal was for an on-site facility, one quite clearly tied to the company's legitimate disposal needs. Because the company's operations in the community were generally deemed beneficial, local residents did not seem to feel it was unfair of the company to initiate a (second) disposal operation within the confines of its Delaware River Plant. Rather, because the company generated the wastes at this location, it made sense for Monsanto to dispose of them there as well, as long as this disposal was conducted safely.

In the SCA case, on the other hand, the issue of equity provided yet another reason for local refusal to accept the proposed landfill. Unlike the Monsanto situation where local waste disposal was directly linked to local waste generation, Bordentown residents felt they were being asked to bear the assorted harms of waste disposal while those who benefited from the initial waste generating activity would have been far removed from the site. Indeed, while the Bordentown landfill was intended to dispose of residual wastes from SCA's hazardous waste treatment and processing facility in Newark, New Jersey, it would also have been open to hazardous wastes generated throughout the state and the region as a whole. Thus, according to the EPA report, "Opponents did not want Bordentown to become the hazardous waste 'dumping ground' for the state."[51]

Overview of Impact-Related Differences. Public response to the Bridgeport and Bordentown landfill proposals, then, was based on considerations of the impacts that these facilities would have on the two communities. The on-site Monsanto facility avoided public controversy largely because it was seen as necessary to the local economy, it was remote from most local residents, it was compatible with existing land use (especially since it was to be located on the ground of the Monsanto plant itself), and the wastes were from local industrial operations and thus linked directly to generation of this material. This facility received easy approval.

In Bordentown, however, the SCA regional (off-site) disposal facility offered no significant economic benefits to the community. The landfill there was seen as being an incompatible, undesirable use of land; local residents did not want hazardous waste disposal to occur opposite the high school, for example. Unlike Bridgeport residents, those in Bordentown were less apt to consider chemical operations to be advantageous for the town or to see the SCA firm as a valuable addition to the local business community. Finally, Bordentown residents also voiced concerns about the fairness of being asked to live with hazardous waste while those who generated the wastes resided elsewhere and derived most of the benefits from its creation. Thus, many in the community felt that they were being asked to bear the risks of these chemicals without receiving commensurate benefits, thereby becoming a "dumping ground." And their opposition prevailed.

Since the SCA-Earthline landfill was defeated during the siting process, there is obviously no way to determine how well it would have operated, or whether community residents would have eventually become accustomed to its presence in their community. Monsanto's on-site landfill met little local opposition, as we have seen, and it was built with significant engineering improvements negotiated during the siting process. This landfill went into operation in August 1978, and for eight months it functioned as designed. In March 1979, however, liquid was discovered flowing into the leak detector area between the primary and secondary liners. Since that date, company officials have been trying, to no avail, to locate the leak. While a 1981 study by Geraghty and Miller consulting engineers found "no firm evidence that the landfill has leaked [into the groundwater],"[52] the company has asked permission to expand the capacity of its landfill. DEP has initiated a comprehensive study of groundwater contamination associated not only with this landfill but with Monsanto's entire complex in Logan Township. DEP, in the meantime, has denied the company's request to expand its landfill capacity further.[53] In this case, even though a hazardous waste facility did not face the obstacle of intense public opposition, it may nevertheless not turn out as an overall success story.

NOTES

1. U.S. Environmental Protection Agency, *Hazardous Waste Facility Siting: A Critical Problem,* SW-865 (Washington, D.C.: U.S. EPA, 1980), p. 1.
2. U.S. Environmental Protection Agency, Office of Water and Waste Management, *Siting of Hazardous Waste Management Facilities and Public Opposition,* SW-809 (Washington, D.C.: U.S. EPA, 1979), p. iii.
3. Michael O'Hare's comments on the draft version of this manuscript helped clarify this point; personal communication from O'Hare to David Morell, 19 January 1982.
4. New Jersey Hazardous Waste Advisory Commission, *Report of the Hazardous Waste Advisory Commission to Governor Brendan Byrne* (Trenton, New Jersey: N.J. Department of Environmental Protection, 1980), p. 1.
5. See Grace Singer, "People and Petrochemicals" in David Morell and Grace Singer, eds., *Refining the Waterfront: Alternative Energy Facility Siting Strategies for Urban Coastal Areas* (Cambridge, MA: Oelgeschlager, Gunn & Hain Publishers, Inc., 1980), pp. 18–63. Michael O'Hare, Debra Sanderson, and Lawrence Bacow, *Facility Siting* (New York: Van Nostrand, forthcoming), Chapter 2, discuss this theme of developer credibility in the controversial hazardous waste facility siting case in Wilsonville, Illinois.
6. Deborah Carmichael, "Hazardous Waste Siting in New Jersey" (Term Paper for Engineering 303, Princeton University, 1981), p. 13.
7. Joyce Gemperlein, "Firm Abandons Plans for N.J. Chemical Dump," *Philadelphia Inquirer* (10 December 1980), p. 7E; Joyce Gemperlein, "Quick Victory Surprises Those Opposing Dump," *Philadelphia Inquirer* (11 December 1980), p. 3B.
8. Donald Janson, "Cost of Hookups Forces Some in Jersey Town To Use Toxic Water," *New York Times* (7 August 1980), p. B2; Michael Brown, "New Jersey Cleans Up Its Pollution Act," *New York Times Magazine* (23 November 1980): 146.
9. James McCarthy, *Public Hearing on S-1300* (Testimony before N.J. Senate Energy and Environment Committee, Newark, New Jersey,

November 6, 1980), pp. 30–33. See Appendix A for a description of the Jackson Township case.

10. U.S. EPA, *Siting of Hazardous Waste Facilities and Public Opposition.*
11. "Judge Sets Trial Date for Stablex Waste Facility Siting," *Hazardous Materials Intelligence Report* (1 May 1981): 3–4.
12. Fox Butterfield, "New England Town Rises Up to Block a Toxic Waste Plant," *New York Times* (17 October 1981): 1, 7.
13. "Local Opposition Threatens IT Expansion into Massachusetts," *Hazardous Materials Intelligence Report* (23 October 1981): 6.
14. Lena Williams, "New York State Finds a Site for Toxic Waste Plant," *New York Times* (13 May 1981): B4. An interesting analysis of the emergence of aggressive public opposition to facility siting from conservative Minnesota farmers is presented in Barry Casper and Paul David Wellstone, *Powerline: The First Battle of America's Energy War* (Amherst: University of Massachusetts Press, 1981).
15. "Carey To Name Special Panel on Toxic Wastes," *New York Times* (27 November 1981): 19.
16. U.S. EPA, *Siting of Hazardous Waste Management Facilities and Public Opposition,* pp. 210–215.
17. Ibid., pp. 303–310.
18. Ibid., pp. 294–299.
19. "Landfill Facility Meets Resistance in Puerto Rico," *Hazardous Materials Intelligence Report* (13 February 1981): 7.
20. Sam Sasnett, *A Toxics Primer* (Washington, D.C.: League of Women Voters Education Fund, 1979), p. 15.
21. This section draws extensively from David Moldenhauer, *Citizen Activists and Hazardous Wastes: A Study in Extra-Bureaucratic Politics* (Senior Thesis in Politics, Princeton University, 1981).
22. U.S. EPA, *Siting of Hazardous Waste Facilities and Public Opposition,* pp. 35–45, 157–175. Our discussion of the Bridgeport and Bordentown cases relies in large measure on this document.
23. Laura Lewis and David Morell, *Nuclear Power and Its Opponents: A New Jersey Case Study* (Princeton, N.J.: Princeton University Center for Energy and Environmental Studies, Report No. 39, 1979), p. 2. The Newbold Island complex was subsequently approved for a more remote location on Delaware Bay in Salem County, adjacent to the Salem Nuclear Generating Station. Here its name was changed to the Hope Creek Nuclear Generating Station.
24. U.S. EPA, *Siting of Hazardous Waste Facilities and Public Opposition,* p. 36.
25. Lewis and Morell, *Nuclear Power and Its Opponents,* pp. 1–2.

26. Monsanto Industrial Chemical Company, Technical Services Department, *Environmental Impact Statement for the Proposed Chemical Landfill, Monsanto Company Property, Logan Township, Gloucester County, New Jersey* (Logan Township, N.J.: Monsanto, July 6, 1977), p. 1. Cited in Peter Montague, *Four Secure Landfills in New Jersey—A Study of the State of the Art in Shallow Burial Waste Disposal Technology* (Princeton, N.J.: Princeton University, Department of Chemical Engineering and Center for Energy and Environmental Studies, draft, February 1982), p. 6.
27. Diane Graves, personal communication to David Morell, 10 February 1982.
28. U.S. EPA, *Siting of Hazardous Waste Facilities and Public Opposition,* p. 37.
29. Montague, *Four Secure Landfills in New Jersey,* p. 8.
30. Graves, personal communication.
31. U.S. EPA, *Siting of Hazardous Waste Facilities and Public Opposition,* p. 39.
32. Ibid., p. 42.
33. Ibid., p. 160.
34. Other HOPE activities included a petition drive and the placement of advertisements in local media. Ibid., pp. 165–166.
35. Ibid., p. 166.
36. Ibid., p. 164. Prior to this action, a bill had been introduced into the state legislature to ban development of hazardous waste facilities near schools, an indirect but not-so-subtle means of disqualifying the SCA landfill proposal. Moreover, both Bordentown Township and Burlington County passed resolutions against the proposed landfill, basing their opposition on an assortment of anticipated negative impacts.
37. Ibid., p. 172.
38. Ibid., p. 166.
39. Ibid., p. 165.
40. "Earthline's attorney was confident that SWA would grant the permit; he dismissed most of the opponents' issues as emotional or specious. The feeling that SWA would grant the permit was shared by HOPE's president and at least some local officials." Ibid., p. 168. In the energy facility siting case studied in detail by Grace Singer ("People and Petrochemicals"), the facility was formally rejected by the state's Natural Resources Council, responding to the strong opposition from local citizens—the mayor and City Council never did act formally on the proposal.
41. See especially Chapter 4 for a discussion of the political problems associated with "easy" preemption.

42. U.S. EPA, *Siting of Hazardous Waste Management Facilities and Public Opposition,* p. 38.
43. See Chapters 3 and 5 for discussion of costs and benefits.
44. Lee Oldham, testimony before the State of California Hearing on "Solving the Hazardous Waste Problem: Non-Toxic Solutions for the 1980's" (Los Angeles, California, 17 November 1980).
45. U.S. EPA, *Siting of Hazardous Waste Management Facilities and Public Opposition,* p. 35.
46. Ibid., p. 168.
47. The siting of the sanitary landfill had been opposed, and landfill operations remained controversial.
48. U.S. EPA, *Siting of Hazardous Waste Management Facilities and Public Opposition,* p. 36.
49. Ibid., p. 43.
50. Ibid., p. 159.
51. Ibid., p. 172.
52. Nicholas Valkenburg and John Isbister, *Ground-Water Investigation in the Vicinity of the New (East) Landfill Monsanto Company, Delaware River Plant, Bridgeport, New Jersey* (Syosset, N.Y.: Geraghty & Miller, January, 1981), p. 5. Cited in Montague, *Four Secure Landfills in New Jersey,* p. 11.
53. Montague, *Four Secure Landfills in New Jersey,* pp. 11–12.

3 WINNERS AND LOSERS: THE SITING PROBLEM IN ITS POLITICAL CONTEXT

The intensity of public opposition to new hazardous waste management facilities, and the consequent inability to achieve success in siting, have led inexorably to growing political pressures to alter the way these siting decisions are being made in the American political system. Should local governments continue to hold veto power over these decisions, or should states reassert their traditional constitutional rights to determine the use of land? In this sense, siting of hazardous waste facilities is fundamentally no different from siting of other controversial facilities needed by society in localities where they are not welcome—jails for dangerous criminals, housing for low-income people, large power plants, storage tanks for liquefied natural gas. This confrontation between majority rule and minority rights rapidly is becoming the most contentious aspect of land-use decision-making, challenging the traditional planning process.

SITING AS AN EXAMPLE OF LAND-USE MANAGEMENT

Controlling the Spillovers

The fact that one person's use of his property may well infringe upon his neighbor's use traditionally has been the cornerstone of efforts to control the manner in which land is used. Nearly three centuries ago, for example, the Massachusetts Bay Colony legislated against operation of "nuisance" industries except in certain designated districts. This action exemplified the modern notion of zoning:

> Local zoning remains essentially what it was from the beginning—simply a process by which residents of a local community examine what people propose to do with their land and decide whether or not they will let them.[1]

Land-use controls emerged as a consequence of society's concern about the *spillover effects* of certain types of land development. Such controls are not required when there is an abundance of land nor when an individual's use of his property does not impact adversely on anyone else. As this frontier situation diminished, however, the possibility that land uses within an area would prove incompatible greatly increased, thus leading to the desire for zoning measures such as the seventeenth century limitation on nuisance industries. In urban areas, in particular, regulation of land use was established to safeguard property values from the actions of others.[2]

Local Control and Home Rule

Control of land is a power vested in the states by the constitution; except on federally owned land, the federal government's direct influence over land management is greatly circumscribed. The states, in turn, began in the 1920s to relinquish nearly all of their power to local communities.[3] In essence, the controversy over the locus of authority to site hazardous waste facilities involves the wisdom of state attempts to regain from the localities their power over large or regionally beneficial new facilities.

The most extreme delegation of local authority, occurring in states like New Jersey and in much of New England, consists of *home rule,* defined as:

> . . . authority granted by either the state constitution or legislature by which municipalities are empowered to set up by local action their form of government and to determine their own substantive and procedural powers.[4]

Local control over land use still has considerable appeal, even in areas with less strict home rule laws. As J. Douglas Peters has noted,

"Implicit in the concept of home rule is faith that local governments can respond sensitively and quickly to local problems, thus fulfilling the ideals of grass-roots democracy."[5] Local management of land use is desired, in part, because residents of a community wish to preserve the fundamental character of their particular community.[6] This theme has strong resonance in American political culture.

Despite localities' substantial political claim to control over decisions on the use of land, such power ultimately remains legally a state perogative. The authority which state legislatures originally delegated to municipalities and counties can be returned to state hands simply by the passage of a new state statute. Townships, villages, cities, and other local communities are, as Judge John Dillon stated in the nineteenth century, mere "creatures of the state"; as such, they have no authority beyond that granted to them by the state.[7]

The Quiet Revolution in Land-Use Control

Over the past decade or so, the preeminence of local control over land use has begun to decline somewhat. The imperatives of state land-use policies, in particular, have served to circumscribe local control:

> It has become increasingly apparent that the local zoning ordinance, virtually the sole means of land-use control in the United States for over a century, has proved woefully inadequate to combat a host of problems of statewide significance, social problems as well as problems involving environmental pollution and destruction of vital ecological systems, which threaten our very existence.[8]

Russell Train, former chairman of the Council on Environmental Quality and former EPA administrator, has suggested one way in which local land-use management is "woefully inadequate": the land-use determinations of one locality may have important consequences for citizens in other areas. In such cases, Train has argued, it is necessary for the broader community affected to have influence over such decisions.[9] This concern is, in essence, the spillover problem writ large. The state's original impetus for establishing a system of local regulation over neighbors' property usage has now been extended to state supervision or influence over the land-management decisions of an entire local community, or even a multimunicipal region. The scale and complexity of modern land development obviously have contributed to this phenomenon. The impacts of a regional shopping mall at the intersection of two interstate highways or of an 1,100-megawatt power plant or a massive oil refinery certainly cannot be restricted to a single municipality.

During the 1970s, states increasingly asserted their rights to decide how land would be used within individual communities. Attorneys Fred Bosselman and David Callies in 1971 labeled this phenomenon a "quiet revolution in land use control":

> The *ancien regime* being overthrown is the feudal system under which the entire pattern of land development has been controlled by thousands of individual local governments, each seeking to maximize its tax base and minimize its social problems, and caring less what happens to all the others.
>
> The tools of the revolution are new laws taking a wide variety of forms but each sharing a common theme—the need to provide some degree of state or regional participation in the major decisions that affect the use of our increasingly limited supply of land.[10]

This concern over a limited supply of land corresponds to the dissolution of the "frontier ethic"[11] in which society does not worry about regulating land use since so much more land is available on the horizon. Yet, as Bosselman and Callies indicate, the American frontier has diminished considerably. This is particularly true in states like New Jersey, which has the highest population density in the nation and in which it is estimated that all of the land with development potential will be in use within the next forty-five to fifty years.[12] The decline of the frontier has brought concomitant changes in attitudes toward land's significance. Traditionally, Americans have viewed property as a commodity—a mechanism by which money can be made. More recently, however, there has emerged a view of land as not only a commodity but as a resource as well.[13] Bosselman and Callies state: "Increasingly, the question being asked is not only, 'will this use reduce the value of the surrounding land?' but 'will this make the best use of our land resources?' "[14]

This expanded conception of land as a resource means that the spillover effects that have been the object of land-use regulation likewise assume a broader character:

> We have experienced a fairly radical change in the way we look at rights in real property—from a philosophy "you can do what you want with your property unless you create a nuisance or violate the zoning laws"

to a recognition that land-use decisions are "affected with broad environmental and aesthetic interests" that may override the owner's right. . . .[15]

The "revolutionary" state laws to which Bosselman and Callies referred are based frequently on this resource-oriented notion of land planning. New Jersey's 1973 Coastal Area Facility Review Act,[16] for example, was designed to provide state oversight over land development adjacent to the state's Atlantic Ocean and Delaware Bay coastlines. This type of legislation was justified on the grounds that local control over land use in this "critical zone" had resulted in a lack of attention to broader regional or state goals. In the absence of state action, this neglect had "unpleasant consequences" for neighboring communities and for the state as a whole.[17] Concern over these consequences illustrates the extent to which attention has shifted away from the concept of land as merely a commodity. With a resource-based view of land, it is possible to see "unpleasant consequences" in any land usage that does not correspond to the "best" one possible.[a]

Determining what might be the "best" means to use a given area is, of course, extremely difficult. It is clear, though, that local planning by itself is not always adequate to this task. One major drawback frequently cited to local control is its inherently limited scope:

> . . . there is an increasing tendency to shift land-use responsibilities from lower to higher levels of government to prevent and repair what are perceived as the undesirable consequences of parochially oriented land-use planning or non-planning.[20]

Despite its parochialism, local control of land use has important advantages. As Peters has indicated, it conforms to a strong Ameri-

[a]The EPA report on public opposition to proposed hazardous waste facilities documents two cases that are particularly good illustrations of how local attitudes toward land use can be strikingly different. In one, the successful siting of hazardous waste disposal facilities in abandoned missile silos in Idaho by Western Containment, Inc. (Wes-Con), the developer was unimpeded by local opposition largely because of residents' attitudes toward land use: "They assumed a traditional attitude in this area that if one owned the land, they were entitled to do with it what they wanted to."[18]

In contrast, the reaction of residents of Rossville, Maryland, to an Allied Chemical Corporation hazardous waste landfill proposal epitomized the resource-oriented view of land described by Bosselman and Callies. Local citizens viewed the Allied proposal as a threat to community land use aspirations: "The community was trying to change its self-image away from the 'dumping grounds of Baltimore County.' Even though Allied promised to restore the brickyard to original grades, the intermediate steps to fill the land (i.e., trucking in hazardous material) did not complement the community's improvement plans."[19]

can belief in grass-roots democracy[21] which emanates directly from the concepts espoused by James Madison and Thomas Jefferson, among others. Likewise, Robert Healy and John Rosenberg have contended that

> . . . local control of land use is, other things being equal, the most desirable arrangement. In judging the merits of most land-use changes, local authorities are not only better informed about the facts of the situation but are also (at least initially) more responsive to the interests affected.[22]

The shift to higher levels of land-use planning encounters many opponents who are mindful of these advantages. "Land-use" 'revolutionaries' have found that home rule, once granted, is guarded tenaciously," noted David Deal.[23] In this respect, the "quiet revolution" that Bosselman and Callies described has not resulted even a decade later in a clear-cut triumph of state over local control; at best, there has been an increasing tendency to look to the states to effect certain types of planning, particularly of facilities with large impacts which are regionally beneficial but are locally unwelcome.

A Not-So-Quiet Revolution

State planning of land use is particularly appropriate in locating such developments of regional benefit.[24] While facilities like airports, prisons, landfills, and public housing produce benefits for an entire state or region, they are frequently opposed by residents of local areas who would suffer the most intense impacts.[25] In these cases, the state's inclination to step in and ensure that these needed facilities are built—for the benefit of the wider community—is checked by a local reluctance to have the state impose an undesirable land use on them, often the type of facility that could otherwise be kept out by use of local zoning authority. As one member of the planning board of a North Carolina county asserted, "We don't want our destiny controlled by the governor. We want to control it ourselves."[26] As a result, state attempts to shift land-use control in these instances of nuisance facility siting have encountered intense and noisy opposition rather than quiet acquiescence:

> Having once delegated most of its authority in land-use regulation to municipal government, the state is regarded as a usurper and intruder

> when it seeks to regain selected aspects of such power and is looked upon as dictatorial when it seeks . . . to exercise supervision over municipal land-use decisionmaking processes.[27]

The stumbling block to state control over certain land uses is not simply a desire by local residents to retain control over their community's land-use destiny. While such local desires and the accompanying political rhetoric are important, an integral part of these state-local tensions in the case of siting hazardous waste facilities must be attributed to the particular distribution of costs and benefits associated with these proposed land uses. The local community is indeed being asked to pay a very high price for the benefit of those people living elsewhere; to many local residents, this simply does not seem fair.

THE COSTS AND BENEFITS OF SITING FACILITIES

The existence of widespread public opposition to proposed hazardous waste facilities can be accounted for by considering the end result of each siting proposal: would the facility be an advantage to the community, or not? In the case of hazardous waste facilities, the outcome is usually perceived locally as far from favorable. That is, from the local community's perspective, the costs of the facility far outweigh its benefits. As a result, most communities that resist specific siting proposals have ample justification for so doing.

Externalities

Ideally, a developer wishing to locate a treatment facility for hazardous wastes would put it where it would do no environmental or health damage (for instance, so far away from people that no one would be exposed to the wastes nor even be inconvenienced by the facility). Unfortunately, no such perfect sites exist anywhere.[28] Even in the remote desert of Nevada or Arizona, or in the remote forests of Maine, the presence of a hazardous waste management facility would impose unavoidable impacts on those few people who do live and work nearby, and on many others who live and work near the routes of transportation for the wastes.[b] The physical range of such impacts depends on the facility and its design. The range of potential effects from an incinerator, for example, is quite large geographically—clearly beyond the radius of just one town.[29]

From an economic standpoint, these "externalities" exist when the social cost of an activity differs from its private cost to the individual conducting the activity.[30] These subsidiary effects can, of course, be either beneficial (positive externalities) or harmful (negative externalities). The existence of negative externalities constitutes a potential reason to oppose the proposed new facility, particularly when these externalities are sufficiently severe.[31] Instances where land uses generate negative externalities are typical of many environmental problems. A factory which emits air pollution, for example, imposes undesired impacts on those who live downwind. Depending upon how injurious these emissions appear, nearby residents might be expected to act to curtail the deleterious effects of this pollution or to deny land-use permits to the facility.

Siting as a "Bad Bargain"

The negative externalities of hazardous waste facilities are so obvious that potential neighbors are highly averse to a facility locating near them, as evidenced by the CEQ survey results noted earlier. The report by Centaur Associates to EPA described the basis for this opposition:

> The community envisions few benefits from the proposed facility—a few jobs and perhaps some tax revenue. Risks are often seen as overwhelming—a "Love Canal" in their community, polluted water supplies threatening the entire community, decades of uncertainty, hundreds of trucks carrying thousands of tons of hazardous waste on local roads.[32]

Most people across the country clearly believe that a hazardous waste facility would force upon them substantial costs and offer them few benefits in return. A 1981 telephone survey of North Carolina residents by Duke University researchers supports this contention. More

[b]That is, even if an ideal site did exist from the viewpoint of remoteness, such a site would be far from ideal in the sense that wastes would have to be transported over a long distance from the industrial areas where they are generated to reach this disposal area, thereby increasing risks to the society at large.

than 50 percent of the survey respondents said they opposed establishment of a hazardous waste facility in their area, even if it would create jobs.[33] In numerous instances, of course, residents of communities facing a hazardous waste facility proposal have considered the costs to be far in excess of the benefits, and have acted accordingly.[c]

The disparity between costs and benefits is true not only of installations that handle hazardous waste but of other "nuisance facilities" as well. Grace Singer noted, for example, that in cases of fuel oil and chemical storage tank siting in New Jersey, citizens in the designated locations saw these proposals as a "bad bargain": "Citizens were concerned that while pollution and safety hazards would be intensified by existence of the facilities, the community would not gain much in return."[36] Importantly, what accounts for a bad bargain in siting is not merely the existence of heavy costs to a community but also the fact that these costs would not be offset by corresponding benefits from hosting the facility.

An essential characteristic of all facility-siting endeavors is the extremely uneven distribution of costs and benefits associated with these projects.[d] The benefits from siting a new hazardous waste facility, for example, are largely captured by its operator, by the waste-generating industries, and by the public-at-large (who presumably will gain from lessened inadequate or illegal disposal of these wastes). The costs, on the other hand, are concentrated within the particular community that hosts the facility, or at best are shared with a few nearby towns.[37] Emergence of local opposition should therefore not

[c]This situation is evidenced not only by the Bordentown rejection of an SCA-Earthline landfill, discussed in the previous chapter, but by several other cases described in the EPA report on public opposition. In Sturbridge, Massachusetts, for example, an attempt by the state Bureau of Solid Waste Disposal to site a secure landfill was opposed by local residents due to its potentially harmful impact on the town's substantial tourist industry which generates $50 million annually for the local economy.[34]

The case of Allied Chemical Company's landfill proposal in Rossville, Maryland, was, in many respects, similar to the Monsanto proposal for Bridgeport, New Jersey, described in Chapter 2. A landfill was sought to dispose of chrome ore waste from Allied's chrome-processing plant in Baltimore; failure to locate a site threatened the continued operation of this plant which employed 350 people. Unlike the Bridgeport residents who did not oppose continuation of Monsanto's chemical operations in the area, however, individuals in Rossville were *not* sensitive to Allied's appeals to the economic importance of its plan. This reaction occurred because the benefits of the Allied operation accrued to Baltimore residents, not those in the outlying area where the company hoped to dump its wastes.[35]

[d]As noted in the following section, "costs" should not be assumed to mean only those factors that can be monetized.

be surprising insofar as the locality is being asked to bear an inordinate share of the proposed facility's overall costs:

> A major impediment to obtaining the cooperation of local communities in siting new facilities is the mismatch between who gets the benefits from the process creating the waste and who has to accept the adverse impacts and risks of taking care of those wastes. This is fundamentally a problem of equitable distribution of benefits and costs.[38]

Costs and Benefits

The costs associated with living near a hazardous waste facility are, without doubt, substantial. An EPA report on compensation and incentives, prepared by Urban Systems Research & Engineering, enumerated a dozen "principal impact issues" (all local costs) that could result from construction and operation of a hazardous waste facility: ground and surface water pollution, air pollution, increased health and safety risk, traffic, noise, odors, aesthetic changes, decreased property values, changes in real estate development, tax revenue impacts, greater burden on public services, and shifts in community image.[39] Michael O'Hare has pointed out that the various kinds of costs imposed on a community frequently may overlap. Fears of health and safety, for example, will be capitalized in changes in property values; or the aesthetic damage of having to look at a hazardous waste facility nearby may reside principally in the fact that the facility's health and safety risk is brought to the viewer's attention every time he sees it.[40] In any event, the scope of potential costs is obviously quite considerable.

Some costs, however, appear to be more severe than others. Thus, instead of the "laundry list" approach to describing costs seen in much of the literature on siting, it may be useful to classify these negative impacts into several types. The most notable categories appear to be:

- health and safety risks;
- nuisance costs and "quality of life" concerns;
- property value and other monetary losses; and
- increased need for community services (depletion of community budget).

These four types of local costs have frequently been the basis for public objections to proposed hazardous waste facilities.[41] Not surprisingly, people's fears often center on the first category, the danger to public health that might result from contamination of the area's water supplies or its air. Part of the opposition to the proposed Allied Chemical Co. landfill in Rossville, Maryland, for example, was attributed to concern that chrome ore wastes would not be contained adequately in the clay but instead would leach out into groundwater.[42] Likewise, the New Jersey Solid Waste Authority's decision to deny a permit for Earthline's proposed landfill in Bordentown was partly due to a judgment that the site chosen would "potentially threaten" air and water quality.[43]

Quality of life and aesthetic concerns can also play an important role in engendering public opposition. The Calabasas landfill operated by the Los Angeles County Sanitation District, for example, encountered opposition from people in an adjacent residential development. Those residents' major concern was aesthetic: the landfill would constitute an undesirable addition to the residents' view of the surrounding hills.[44]

Land-use compatibility is another important aspect affecting quality of life. The SCA experience in Bordentown, described in Chapter 2, demonstrated how this issue can play a part in establishing local opposition to a facility. Likewise, community perception of appropriate land use was instrumental in the Minnesota Pollution Control Agency's rejection of a demonstration chemical landfill facility. This particular siting process encountered problems in establishing acceptable siting criteria to be used in choosing a location for the proposed landfill. A fundamental objection to the criteria that were offered was that they did not show sufficient concern for preserving the agricultural integrity of various rural regions.[45] The EPA report states:

> Use of agricultural land for other purposes has become a political issue in Minnesota, particularly development which is perceived as being largely for the benefit of urban residents (e.g., high voltage power lines). Rural residents were opposed not only to the facility itself but also to the industrial development which it might have attracted.[46]

The issue of land-use compatibility need not, of course, be a factor that militates against siting of a facility, as evidenced by the Monsanto landfill siting in Bridgeport, N.J. The successful siting of a hazardous waste treatment facility by Frontier Chemical Waste Proccessors, Inc., in Niagara Falls, N.Y., was also in large measure due to its compatibility with existing land use:

> The site had been used for manufacturing chemicals for 60 years and was in an industrial area; Frontier's operations would then conform to adjacent uses, be relatively inconspicuous and be consistent with previous activities at the site.[47]

Not surprisingly, public reaction to the Frontier waste treatment facility was virtually nonexistent inasmuch as the facility did little to change the basic character of the area in which it was sited.

A somewhat more tangible concern about the costs of hazardous waste facilities is that siting will result in a decline in property values.[e] The resistance to the Stablex Corporation's proposed waste-solidification facility in Hooksett, New Hampshire, for example, has been partly based on the town planning board's contention that the facility under dispute would result in reduced property values and thereby erode the town's tax base.[49]

Moreover, opponents of hazardous waste facilities argue that the additional costs to a community that hosts such a facility make it a financial liability to a town. One of many issues apparent in the unsuccessful siting process initiated by Minnesota's Pollution Control Agency was the anticipated result that a hazardous waste management facility would drain a town's budget by its demands for local services.[50] In testimony to the New Jersey Senate Energy and Environment Committee during its revision of the state's siting legislation, several local officials from different communities expressed concern about the adverse effect of hazardous waste facilities on municipal finances, particularly in terms of providing adequate emergency response capabilities. Such concerns prompted New Jersey to establish a gross receipts tax to repay host communities for the direct financial impacts of these facilities.

[e]The effect of hazardous waste facilities on property values need not, of course, be a deleterious one. Some argue, for example, that inasmuch as these facilities represent an additional tax ratable for a town, property taxes in a host community can be lowered, thereby *increasing* property values relative to the rate of taxation. Hence, the effect of a hazardous waste facility on property values is ambiguous. Should the facility experience serious environmental problems, however, it is certain that property values would plummet, as was the case at the Love Canal in Niagara Falls, New York.

An interesting aspect of this question of the effect on property values of a hazardous waste facility siting was the reaction of some Missouri residents to a 1977 landfill siting by Bob's Home Service. Surprisingly, those residents closest to the proposed landfill did *not* oppose the landfill, unlike others in the area, for fear that adverse publicity would endanger property values.[48]

In contrast to the costs of hosting a hazardous waste facility, the benefits to the community itself are rather limited. Most obvious are increased tax revenues and the provision of (a few) jobs. Employment opportunities associated with construction and operation of hazardous waste facilities are not a major attraction. In the siting attempt in Starr County, Texas, for example, the benefits used to try to sell the proposed landfill to the county involved the creation of *four jobs* on the site as well as "many" new opportunities for waste haulers. Not surprisingly, these benefits did not serve to counteract the strong local opposition to the proposed facility.[51] A $10 million facility for solidification/stabilization of hazardous wastes proposed for construction in Hooksett, New Hampshire, would result in employment of only thirty to thirty-five people, about half of whom would be previous residents of the community.[52]

Tax benefits generally prove to be a major attraction for local communities. In recent years, various states have passed different statutes allowing development of hazardous waste facilities. Some of these state laws require the developer to make payments in lieu of (or in addition to) normal taxes on real property. This may include paying the host community a stipulated percentage of the facility's gross receipts. For example, California's S.B. 501 authorizes the local host community to tax a facility up to 10 percent of its gross revenues; New Jersey's S. 1300 authorizes a 5 percent gross receipts tax; Georgia law mandates a 1 percent tax.

Data on the gross receipts of specific hazardous waste facilities are not readily available, and they certainly vary greatly from one facility to the next. According to Paul Abernathy of Chemical Waste Management, Inc. (CWM), it is very rare for a single site in the hazardous waste industry to gross over $10 million per year.[53] Perhaps two or three sites in the United States are of this scale. One is the BKK facility in West Covina, California, which handles 35 to 38 percent of all the hazardous wastes in California. With estimated annual gross receipts of $10 million, the facility pays about $1 million in taxes to the local community.

CWM is one of the largest firms of its kind in the United States. It grosses about $125 million per year from hazardous waste management at its eighteen sites. While these facilities vary in size, their average gross receipts are some $6.9 million annually. Tax payments from these facilities vary from one state to another. For communities facing new siting proposals for hazardous waste facilities, the following data based on the CWM national average may provide a useful guide:

Gross Receipts	*1% Tax*	*5% Tax*	*10% Tax*
$6.9 million	$69,000	$345,000	$690,000

Another set of estimates was provided by Larry McCoy of the Stablex Corporation, based on their proposed facility in Hookset, New Hampshire. New Hampshire requires that a fee of $6 per ton be paid to the host community. Since this company's proposed Hooksett facility would handle 100,000 tons per year, the result would be a payment of $600,000 per year. The facility would also pay property taxes, based on a rate of $29 per $1000 of assessed value. The proposed facility would cost about $10 million in capital expenditures, generating annual property taxes of some $300,000.[54]

Obviously, facilities are less likely to be perceived as a bad bargain by those communities that most need property tax ratables, and which already have significant industrial land uses. In New Jersey, this situation describes Newark, a city which, unlike other less industrialized communities in the state, has expressed some formal interest in hosting a new hazardous waste management facility. Indeed, the mayor's office has actively sought new facilities. SCA Services already operates a hazardous waste facility in Newark, and At-Sea Incineration, Inc. plans to build a storage and transfer facility in Port Newark, on a portion of the waterfront to be leased from the Port Authority of New York and New Jersey.[55] Even in impoverished Newark, however, citizen resistance to new waste facilities has been evident, based primarily on residents' fears. Residents of Newark's Ironbound section have been particularly opposed to this siting proposal.[56] Hazardous waste facilities might also be seen as beneficial insofar as they generate a demand for private services and support other local industries that require disposal capacity.[57] Taken together, however, all of these benefits usually do not add up to very much from the local perspective, and the emergence of local opposition ought not to be surprising.

The foregoing discussion has referred to costs and benefits as wide-ranging kinds of impacts whose effects are perceived as either harmful or beneficial. Importantly, the costs and benefits associated with siting, particularly costs, are not confined to impacts that can be readily quantified. A report prepared for the state of New Jersey and the Delaware River Basin Commission by the Booz, Allen & Hamilton consulting firm noted that:

> One of the main concerns of the public about waste management operations is the potential community impact and cost of such facilities. These

costs can be expressed quantitatively in terms of lowered property values, or as costs of necessary tax-supported services such as fire protection or road maintenance. Costs may also be noted in more subjective terms, such as in the form of nuisances to local residents (e.g., odors or general aesthetic unpleasantness), or "damages to community values" which would be expected to occur if the community became known as a regional center for waste disposal.[58]

In the past, such unquantifiable costs have often been ignored, with attention focused on only the most tangible costs and benefits. As Ronald Luke has noted, prior to the 1970s community acceptance of facility siting was extremely high: "For most communities, a new major facility meant economic and population growth, which was viewed as an unqualified benefit."[59] To the extent that the benefits of facility siting can be readily quantified while the costs remain far less tangible, a significant asymmetry exists. Traditionally, this has meant that benefits have had an inherent advantage in the balancing process. This advantage arises out of what David Bradford and Harold Feiveson have referred to as a sense of "legitimate discourse" in a political community's policymaking:

> For a variety of reasons, certain types of arguments or values cannot easily be conveyed or invoked. Thus politicians will be more likely to stress hard data (such as construction costs and tax benefits) than they will concepts such as beauty or solitude. In this sense intangibles are at a disadvantage.[60]

Nevertheless, the increased environmental consciousness of the 1970s has brought about attempts to take these unquantified costs into account. Litigation under the National Environmental Policy Act (NEPA), for example, prompted a 1971 federal Court of Appeals' ruling, *Calvert Cliffs Coordinating Committee v. AEC,* that unquantified costs must also enter into the decisionmaking process:

> In order to include all possible environmental factors in the decisional equation, agencies must identify and develop methods and procedures which will ensure that presently unquantified environmental amenities and values be given appropriate consideration in decisionmaking along with economic and technical considerations.[61]

Michael Baram has devised a framework to consider both quantifiable and unquantifiable impacts. He classified externalities that result from construction and operation of a facility as follows:

- *Ecology:* includes erosion, landscape changes, and air and water pollution.
- *Economy:* includes private property values, taxes, local and regional jobs.
- *Regional and community quality of life:* includes human health (both physical and psychological), aesthetics, congestion, odors, traffic.
- *Social and political factors:* includes changes in residents and life styles, changes in social opportunities, changes in municipal systems.[62]

Examination of these categories of impacts reinforces the conclusion that from the local community's perspective, in most instances, the costs of having a hazardous waste management facility greatly outweigh the benefits. The benefits that accrue to a community from the siting of a facility are largely economic and, as just noted, relatively slight. In contrast, the potential costs are found in all four categories, the most important, perhaps, being the adverse effect on "regional and community quality of life," including fear about public health risks.

Some impacts, though, are more certain than others. The siting of hazardous waste facilities involves undesirable occurrences which, in actual practice, may never occur; even the possibility of these events taking place, however, is rightly counted as a cost. Some of these costs can be quantified through probability analysis, although this is very difficult and very controversial. A crucial question when balancing costs and benefits thus becomes whether or not the risks are "worth it" to the local community and to the broader society.

The Politics of Acceptable Risk

One of the basic principles of facility siting is that some risk is unavoidable, especially to the residents of the area immediately around the facility. "Even with the best standards, technology, management, and enforcement, each hazardous waste management site will have some residual risk," the Keystone Center conferees concluded.[63] The inevitability of risk does not mean, however, that these facilities—especially the modern generation of waste management facilities—are necessarily unsafe. Indeed, whether or not a project is safe is not an inherent function of its design and operation; rather, assessments of safety entail profound value judgments by all of the individuals involved. As William Lowrance put it:

> Much of the widespread confusion about the nature of safety decisions would be dispelled if the meaning of the term *safety* were clarified. For a concept so deeply rooted in both technical and popular usage, safety has remained dismayingly ill-defined.
>
> We will define safety as a judgment of the acceptability of risk, and risk, in turn, as a measure of the probability and severity of harm to human health.
>
> A thing is safe if its risks are judged to be acceptable.[64]

The crucial question in dealing with a given risk thus becomes whether or not it is deemed acceptable, and by whom. These calculations of whether a particular risk is acceptable or not are set against judgments about the corresponding benefits to be obtained by taking these risks. Studies have shown that acceptance of risk increases when the benefits to be derived from an activity also increase.[65] That is, each individual assesses the available package of risks and costs and corresponding benefits (jobs, income, lower taxes, and so on). One particular combination might be acceptable (or tolerable) to some people, but not to others—depending on their value preferences, available options and alternatives, and so on. And that calculation might well change over time for some of these people, as their own circumstances altered in some important way. In this sense, acceptability is not a simple property that can be predicated on a particular risk. According to Michael O'Hare, you can find people who will accept almost any risk over a very broad range if you put it in the right package, and also people who will not accept the same risk if you put it in the wrong package.[66]

Numerous other considerations may enter into judgments of the acceptability of risk. One of the most important is whether the risk is assumed voluntarily and consciously (even if one's available alternatives are few), or is involuntary. Studies of risk acceptance have shown, for example, that individuals will accept voluntary risks roughly 1,000 times greater than involuntary exposure risks.[67] The reasons for this differential are quite obvious: risks are assumed voluntarily when corresponding benefits can be consciously considered to be worth the risk. Involuntary risk, in contrast, offers no such apparent compensatory benefits to the individual involved.[68] Such risks typically are imposed on one person by another.

The voluntary versus involuntary distinction in risk-benefit calculation is therefore akin to the bad bargain phenomenon in facility siting. To local residents the costs associated with the imposed risk appear to outweigh the benefits, perhaps in large measure because the benefits offered are not necessarily desired (or needed) locally. They certainly are not sufficient to offset the imposed new risk.

A rational calculus of whether the benefits of an activity justify the risks incurred from this activity becomes extremely difficult, however, in the case of hazardous waste facility siting. Unlike the case of automobile use, where the average American each year runs a risk of about 1 in 4,000 of dying in an accident,[69] the actual risks associated with a hazardous waste facility over its years of operation are extremely uncertain. It is well known that past facilities like the Chemical Control Corporation in Elizabeth, New Jersey, have offered inordinate risks; but how risky are new "legitimate" waste treatment facilities? We simply do not know.[70] Judgments of the safety of new, high-technology hazardous waste management facilities range from "environmentally safe" to "a black art," depending on the values of the observer.

In debates on siting hazardous waste facilities, "technical arguments can be found to support or refute any conclusion offered."[71] Much of this technical uncertainty can be attributed to a lack of data and the lack of operating experience. Unlike automobile fatalities or airplane crashes, there is virtually no track record for legitimate hazardous waste facilities;[72] licensing requirements under RCRA began to apply only in November 1980. Indeed, without final implementation of these regulations or knowledge of how stringently they will really be enforced, deducing the true risks to the host community of a hazardous waste facility requires considerable guesswork.

Yet uncertainty over risks would still be present even if a longer operational history were available for hazardous waste facilities of the same type called for in recent siting proposals. One important topic in risk calculation is the probability of extremely improbable events; when the probability of occurrence within a specified time is extremely small (for instance, "one in a million"), there is again an obvious problem of the lack of experience or data. In the case of a serious accident involving a nuclear reactor, for example, Alvin Weinberg stated (several years before the accident at Three Mile Island), "Because the probability is so small there is no practical possibility of determining this failure rate directly—i.e., by building, let us say, 1,000 reactors, operating them for 10,000 years and tabulating their operating histories.''[73] A similar situation may apply for certain types of hazardous waste management facilities.

In this regard, moral considerations become evident—certain risks are so large or so terrible that they really become unacceptable in the broad sense. O'Hare notes that most people would feel guilty if they negotiated a price at which they would accept such risks, even when they might accept them implicitly as part of a package presented in a different way. A large part of the politics of siting hazardous waste facilities has to do with persuading people that while the risks of hazardous waste disposal are real, they are not of the unthinkable or immoral category of risk but in the thinkable, livable category of life's risks such as living near an ordinary factory or a chemical plant.[74]

In the face of all this uncertainty, risk acceptance decisions become particularly susceptible to differing sociopolitical values. A risk-averse view holds that in dealing with hazardous substances under uncertainty about their long-term effects on human health, ''restraint is wise.''[75] This view would lead people to deem unsafe even slight risks of exposure to hazardous substances. In contrast, of course, individuals more optimistic about the efficacy and safety of modern waste treatment and disposal technologies are more likely to accept the (small) risks associated with operation of such hazardous waste management facilities.[76] Nearly everyone in the hazardous waste treatment industry holds the latter perception, one which is shared by many officials of the chemical industry and by many in state regulatory agencies. In contrast, many environmentalists fall into the first category, as do many residents of those communities actually selected as the site for a new waste facility.

Another way to judge the ''reasonableness'' of uncertain risks is to try to compare them to accustomed practice:

> The underlying assumption is that if the thing has been in common use it must be okay, since any adverse effects would have become evident, and that a thing sanctioned by custom is safer than one not tested at all.[77]

This attitude toward risk has important implications for siting hazardous waste facilities. Many believe that these facilities can be designed and operated to be no more risky than other industrial processes. For example, Leonard Tinnan of Chemical Waste Management, Inc., testified at a 1980 California hearing that:

> With closed vapor control storage tanks and other appropriate provisions, a hazardous waste transfer station is literally no different in appearance or operating mode than the many hundreds of petroleum and chemical storage facilities that now exist throughout California. If properly constructed and operated, and if located in suitable industrial zone areas, transfer stations yield no adverse environmental impact.[78]

And John Theiss of the Industrial Tank (IT) Corporation said succinctly at the same hearing: "A technologically advanced waste management facility is like any other industrial or manufacturing facility."[79] The logic behind this argument is that because other industrial processes are thought safe, new-generation hazardous waste facilities should also be considered safe since their risks are comparable. A report prepared for the New England Regional Commission noted:

> . . . many petrochemical and other heavy industries have the same types of toxic materials as inputs as do hazardous waste management facilities, yet communities are not alarmed by them. Every industrial process poses risks, each to varying degrees. . . . A resource recovery facility can release toxic fumes and sludge. A coal or oil-fired plant poses well-known, yet accepted, risks.[80]

This type of reasoning, however, might just as easily run the opposite way. That is, if hazardous waste facilities are adjudged to have unacceptable risks, then other industrial facilities with similar risks should also be considered unsafe to the surrounding public.[f]

Public consideration of unknown risks in siting (which leads, most often, to local opposition to proposed facilities) has, according to some, been based on an irrational approach. A draft report by Urban Systems Research & Engineering maintained that:

> While public opposition often centers on identifiable potential impacts of hazardous waste management facilities, it is clearly not always a rational process. The unknown risks associated with hazardous waste management facilities often create an attitude of fear and mistrust. The public emphasizes the uncertainty of the risks, and questions the ability of government, industry officials or *anyone* to assure long-term safety.[81]

[f]It is quite likely, in fact, that the true risks for other industrial facilities have not been known, much like the risks of hazardous waste facilities were not widely known until the late 1970s. In this case, a siting policy for petrochemical plants or resource recovery facilities may ultimately prove necessary.

The attitude of fear and mistrust referred to here ought *not* to be considered irrational, however. Though the risks associated with a new facility are uncertain, past experience with hazardous waste facilities—especially landfills—clearly justifies a view that serious risks to the public are associated with these wastes. Unless sufficient evidence can be presented to show that these previous conditions would not hold in the present period,[82] public opposition seems quite justified. It is thus eminently rational for people not to want a new chemical waste facility in their backyard. These fears will justifiably persist until there is ample reason to believe that new facilities will not prove to be dangerous "chemical dumps." Each individual's assessment of probabilities and the value of consequences is intrinsically subjective. There is certainly no reason to expect everyone in a community to assess the probabilities of risk or to value proposed benefits in the same way. As a result, there will always be some local opposition in any community proposed as the site of a new hazardous waste facility.

This situation suggests an important relationship between knowledge and risk perception. Before the dangers of hazardous chemicals—and the horrible experiences of many of those exposed to them—were widely publicized, the perceived risk of a hazardous waste facility was undoubtedly much less. Although the actual risk of the facility could well have stayed the same,[83] perceptions of that risk and the corresponding assessment of its acceptability will differ depending upon an individual's level of knowledge. To the extent that citizens are now "overly" fearful of hazardous waste, as the chemical industry has argued in its campaign against "chemicalphobia,"[84] future siting for the new generation of hazardous waste management facilities may have to reconcile the difference between the perceived risks associated with the facility and more objective technical assessments of this risk.

> Some people's exaggerated perceptions about risk, however, may be the most serious obstacle to successful siting of new facilities. The way information about risk is conveyed to the public, therefore, is crucial.[85]

One of the recommendations of New Jersey's Hazardous Waste Advisory Commission was, therefore, to establish a public information program so that individuals might better be able to assess risks:

> A public education program is needed to advise the public of the facts about hazardous waste, including any health risks, so that they can make informed judgments about different kinds of facilities that may be proposed.

> Education must not be confused with propaganda, however, nor will education be an antidote to public opposition. There are identifiable (though not necessarily attributable) risks associated with hazardous waste treatment and disposal facilities, and the public's concern about them must be addressed and resolved.[86]

Despite efforts to educate people about the risks involved in a facility siting proposal, decisions as to a particular facility's acceptability will continue to demand value judgments about the relative balance of risks and benefits. In this sense, such decisions may be less conducive to public education. Industry will, of course, try to convince the public to be amenable to what it considers "sensible solutions,"[87] but these will largely be attempts at political rather than technical persuasion. Although risk assessments frequently revolve around technical issues, decisions about "acceptable risk" are thoroughly political. Discussion of risk thus constitutes what Alvin Weinberg has termed a "trans-scientific" problem:

> Many of the issues which arise in the course of the interaction between science and technology and society—e.g., the deleterious side effects of technology, or the attempts to deal with social problems through the procedures of science—hang on the answers to questions *which cannot be answered by science.* I propose the term *trans-scientific* for these questions since, though they . . . can be stated in the language of science, they are unanswerable by science; they transcend science.[88]

To call judgments about risk acceptability and safety "trans-scientific" recognizes that individuals' values and political beliefs will differ. It should not be surprising, therefore, that when these divergent views are brought forth in response to actual hazardous waste facility siting proposals, intense conflicts arise.

IN SEARCH OF THE PUBLIC INTEREST

Winners and Losers in Siting

A necessary component of environmental decisionmaking, according to R.D. Smalley, includes consideration of who wins and who loses:

> . . . we can have all the risk assessments in the world, and they won't matter one whit if it is not agreed in the first place who ought to take the risks and who ought to get the benefits from the project producing the risks.[89]

In the case of hazardous waste facility siting, though, the mismatch between costs and benefits is acute. Residents of a community where a facility is proposed see themselves completely as potential losers, inasmuch as siting offers them a bad bargain. Indeed, the individual community where a facility would be sited may well stand to lose.

Nevertheless, there appears sufficient reason to devise effective state siting policies. Ironically, states like New Jersey to a large extent already have "sites" for the wastes being disposed: in the absence of planned siting by the state, the siting is done all too often by illegal disposal.[90] In such cases, residents of communities across the state incur substantial health costs while obtaining absolutely no offsetting benefits.[91] Thus, while most residents of a community facing a proposal to locate a new hazardous waste management facility in their midst may deem the risks unacceptable, important risks must be borne by the state as a whole if the facility is not sited. Paradoxically, then, risk-averse decisions of the type made by individual local communities carry special risks of their own for the broader statewide community.[92]

Uneven distribution of costs and benefits—local assumption of costs versus statewide receipt of benefits—constitutes only part of the facility-siting paradigm. Also notable is the fact that the benefits of siting new hazardous waste facilities are dispersed, whereas the costs are highly concentrated. To some extent, the benefits accrue to each individual in the state whose risk of exposure to hazardous substances is significantly lessened. While the per capita benefit may be low, when these health benefits are summed the total could be substantial. This may depend, of course, on the specific situation. Benefits may flow to people who use the products for which hazardous waste disposal is made less expensive than shipping it long distances, or to certain taxpayers who would otherwise have to pay for eventual cleanup operations. And there are, of course, large per capita benefits to select individuals and groups, especially to the developer and certain corporations which use the facility. In any event, the benefits are essentially dispersed, especially in contrast to the costs (for example, risk to health and other quality-of-life considerations) which are distributed among a much smaller group: local community residents, and perhaps the residents of adjoining communities. Importantly, the impact on each person among this smaller group is felt to a much larger extent. As a result, although the total benefits from a new facility may outweigh its total costs, the average per capita costs easily exceed the average per capita benefits.[93] This latter situation bears heavily on the political viability of siting proposals.

Zero-Sum Siting and the Policy Impasse

If the interest of the state as a whole lies in siting new facilities, what prevents states from guaranteeing that siting will occur in order to bring about this benefit? Some states, of course, *have* acted, frequently opting for preemptive measures, as described in Chapter 4. The majority, however, have *not* initiated state-run siting policies for new hazardous waste management facilities. The explanation lies partly in political inertia, but more intriguingly in the political system's bias against allocating without due compensation large losses to identifiable small components of society, a bias traditional in American political culture.

In *The Zero-Sum Society,* Lester Thurow has pointed out that desirable policies frequently cannot be carried out due to a political reluctance to impose clear economic losses on specific groups. In one sense, his characterization of this problem aptly fits the siting case, except that the losses imposed by siting are far more than simply economic ones:

> For most of our problems there are several solutions. But all of these solutions have the characteristic that someone must suffer large economic losses. No one wants to volunteer for this role, and we have a political process that is incapable of forcing anyone to shoulder this burden. Everyone wants someone else to suffer the necessary economic losses, and as a consequence none of the possible solutions can be adapted. . . .
>
> When there are large losses to be allocated, any economic decision has a large zero-sum element. The economic gains may exceed the losses but the losses are so large as to negate a very substantial fraction of the gains. What is more important, the gains and losses are not allocated to the same individuals or groups. . . .
>
> The problem with zero-sum games is that the essence of problemsolving is loss allocation. But this is precisely what our political process is least capable of doing. When there are economic gains to be allocated, our political process can allocate them. When there are large economic losses to be allocated, our political process is paralyzed.[94]

While the United States has not devised sophisticated compensation mechanisms as has been done in Japan for certain actions (such as sulfur oxide air pollution), certain public decisions that do impose focused costs *could* be made in conjunction with mechanisms that reallocated the benefits along the same pattern. In the case of siting hazardous waste facilities, Thurow's analysis illustrates why—absent effective compensation—communities often respond to proposals with the view: "We don't object to the idea of a hazardous waste facility as long as you build it someplace else."[95] This attitude on the part of potential losers is quite understandable. Presumably, however, facility siting is *not* a zero-sum activity; if the facility is worth building, there is a net benefit to the decision to build it. And this net benefit could conceivably be captured for reallocation to those who would otherwise suffer unacceptable costs.

While local resistance to siting proposals, without adequate compensation, is readily understandable, less evident is why the political process is, as Thurow asserts, "paralyzed" to the point that states find it difficult or impossible to effect siting policies for the larger benefit. One explanation for this paralysis may be a preference for the status quo by political decision centers. Urban planner Edmund Burke has explained this preference as follows:

> A community decision is usually the consequence of competing and conflicting claims and values. To approve one decision will please one group and displease another. [Decisionmaking] involves hard choices in which there may be winners and losers. Decisionmakers can be expected to shrink from such choices.[96]

The political costs and benefits of siting for decisionmakers may run counter to the overall cost-benefit configuration imposed by the facilities themselves. Although net benefits may be obtained by establishing sites, the harm is especially apparent to those who stand to lose by such actions.[97] Moreover, since the per capita costs are much greater than the per capita benefits, the losers are likely to make themselves heard politically, while the winners (except for the facility developers) do not have as much at stake and thus are far less likely to exert the same political pressure for their own cause.

The interests of low-win beneficiaries may not enjoy proportionate weight in the political arena partly because high-loss "victims" of siting have an important organizational advantage:

> . . . neighbors for whom the [facility] is costly on net are likely to invest significant effort in opposing it, while the more diffuse groups of beneficiaries are likely to remain inert, reflecting the rational expectations of each of their members that his own action will not affect the result that in fact ensues.[98]

This problem is compounded, of course, when the potential beneficiaries are not even aware of their gains should a certain policy be pursued. This is certainly the case in siting, since most state residents are not even aware that hazardous waste facility siting policies would offer them any benefit. It should not be surprising, then, that the broader social need for new hazardous waste facilities has not been well represented in the siting process, and that opposition to siting from local citizens has tended to dominate these decisions.[99]

The Public Versus the Private Interest

According to Michael O'Hare, "The siting problem should be viewed as a special case of the problem of resource allocation between small but concentrated and large but diffuse interests."[100] Given America's political institutions, however, such allocation poses an enormous challenge. This is especially true in the case of siting hazardous waste facilities, in which the "small but concentrated interests" may—like the individual in Warrenton County, North Carolina, who vehemently opposed PCB siting—perceive the outcome of this allocation as a matter of life and death.

Thurow has argued that politics in the United States operates on the premise that governmental action will not impose losses. As a result, "In our system, proposals that yield economic losses come as a surprise, are treated as a betrayal, and result in fierce political resistance that makes it impossible to impose the programs."[101] The hesitancy of American political institutions to allocate losses is not a universal phenomenon. In France, for example, the judicial system has not been sympathetic to opposition to nuclear power insofar as these opponents are viewed as opposing the "public interest."[102] American institutions, with their Jeffersonian heritage and decentralist bias, are clearly more tolerant of the "private interest."

The larger public interest does not, of course, always go unserved. State land-use innovations in recent years have tended, as previously noted, to allow state intervention in cases where a particular land use affects not only a locality but the state as a whole. In Massachusetts, for example, the state's interest in provision of low-income housing prompted passage of "anti-snob zoning" legislation. In granting approval for this measure, a Massachusetts court suggested broad sanction for such public interest measures:

> The strictly local interests of the town must yield if it appears that they are probably in conflict with the general interests of the public at large, and in such instances the interest of the municipality would not be allowed to stand in the way.[103]

Much like the override of exclusionary zoning,[104] siting of hazardous waste management facilities is very much an issue of state gain versus local harm. Such endeavors encounter immediate resistance to the extent that they infringe upon local control. In the debate over siting legislation in New Jersey (S-1300) the initial stance of many local officials was: "Our home rule is being eroded."[105] It remains to be seen, then, whether the states can effectively advance the broader public interest in the face of such local resistance, or whether blunt attempts at preemption will simply intensify citizen anger and make the situation worse rather than better.

The Crossroads of Equity and Efficiency

Local communities facing the enormous costs associated with imposition of a hazardous waste facility are unlikely to be very impressed by arguments in favor of state actions in the public interest. State gain may justify local harm in some instances, but what if the harm is prohibitive? In other words, there is undoubtedly some level of local harm at which the overall state benefit can no longer be asserted as the cause to impose the facility over local opposition. This concern over *equity* is a fundamental aspect of all the debates over siting. While decisionmakers may aim at achieving an efficiency goal of establishing a policy that results in the greatest net benefit, equity considerations must feature prominently in such decisions as well. In Frank Michelman's analysis:

> There is, after all, strong appeal in Hobhouse's insistence that a rational social order does not rest "the essential indispensable condition of the happiness of one man on the unavoidable misery of another, the happiness

> of forty millions of men on the misery of one''; and it is not easy to disagree with his statement that, however temporarily expedient, ''it is eternally unjust that one man should die . . . for the people.''[106]

Considerations of alternative siting policies must therefore not only inquire about their efficiency (like the Massachusetts Court's concern for ''the general interest of the public'') but also about their fairness. In fact, perceptions of political equity may be very important contributory factors in engendering local opposition to a new hazardous waste disposal site. As indicated in Chapter 2, for example, residents of Bordentown, New Jersey, were highly averse to becoming a ''dumping ground'' for hazardous wastes generated elsewhere in the state. People do not want to feel exploited, misused, or taken advantage of; concerns over equity may therefore dominate their response to a facility-siting proposal. The Duke University telephone poll of North Carolina residents illustrates how the issue of geopolitical exploitation can combine with the prevalent ''not-in-my-backyard'' attitude: only 7 percent of the respondents said they approved of allowing wastes generated out of state to be disposed of in their county. About 38 percent said they would be willing to allow wastes from within the state, and 49 percent said they would only accept wastes from within their county.[107]

Perhaps the most striking example of citizen opposition to a proposed hazardous waste management facility based on political equity comes from Starr County, Texas, an area near the Mexican border. In this case, Starr Industrial Services, an outside firm, attempted to obtain approval for a technically superb hazardous waste landfill. Local residents recognized that the risks presented by the site were negligible, but they resisted the facility nonetheless. As a member of one women's group wrote to the Texas Water Quality Board, ''Our soil may be impermeable, and the chances of waste products escaping into our underground water may be slight, but we do not want this hazard here, however slim this hazard may be described.''[108] The primary reason given for this resistance is that Starr County residents did not want a facility in their area to dispose of wastes produced outside the county. The question of equity was not contingent upon the risks the facility posed, but rather fed upon pre-existing feelings of alienation and injustice. Being poor and predominantly Mexican-American, residents of Starr County had traditionally been exploited by wealthier, white-dominated areas of the state. The wastes planned for the Starr County facility originated in Houston's petrochemical complexes; sending them to Starr County emphasized local residents' feelings of alienation:

> Starr County, in the minds of its inhabitants, has the image of being a community of poor, second class citizens that is generally being "dumped on." . . . Starr County inhabitants have an average per capita income that is approximately one-third of the state and national average. The elected officials as well as those trying to increase the economic prosperity of the area saw the introduction of a hazardous waste facility as contributing to this bad image and as a further sign of being dumped on.[109]

The growth of resistance in Starr County was therefore less the result of perceptions of inordinate risk common to most hazardous waste facility-siting attempts, and more the feeling that the county was being exploited as the site of this facility. The introduction of hazardous wastes into the area symbolized the economic and social exploitation against which the county's residents were beginning to rebel.

As in Starr County, municipalities in which sites are proposed are likely to feel that they are being exploited for the benefit of the state or broader region, and to reject such policies as "domestic imperialism."[110] Yet either a certain amount of exploitation, or highly innovative programs of compensation, may well be necessary if any net gains are to be made. The alternative, as Thurow indicates, is public policy paralysis—foregoing beneficial policies entirely because some people will endure losses.

Ultimately, then, a difficult tradeoff is required. According to William Murray and Carl Seneker, ". . . siting becomes a problem of finding the best way of using available resources to maximize the quality of life. It is basically a question of tradeoffs."[111] In making these tradeoffs, conflicts are likely to be manifest. The coercive element involved in siting cannot be overlooked. To the extent that a state "foists" hazardous waste facilities on an unwilling community, it is also imposing involuntary risks that local residents are likely to deem unacceptable. Unlike Garrett Hardin's "Tragedy of the Commons" in which the solution to overexploitation of a common resource consists of "mutual coercion mutually agreed upon,"[112] the siting case typically involves state coercion strongly resisted locally. Hardin's solution requires that everyone share the costs of producing a new social benefit; siting policies like New Jersey's S.1300 require that only *some* bear the costs to produce this benefit. Deciding who these losers will be and how they will pay is a necessary consequence of any coercive (preemptive) siting plan.[g]

Thus, the siting problem fundamentally involves political choices. The tradeoffs between equity and efficiency, for example, clearly rest on value judgments.[h] As Bradford and Feiveson cautioned:

> Because alternative policies will typically have important distributional effects in which some people will gain and others will lose, the decisions are intrinsically political and they must be decided through the political process.[112]

In a process fraught with conflicts, the extent to which siting policies can come to terms with these difficult political questions will certainly be a major determinant of their eventual success.

NOTES

1. Fred Bosselman and David Callies, *The Quiet Revolution in Land Use Control: Summary Report,* Prepared for the Council on Environmental Quality (Washington, D.C.: U.S. Government Printing Office, 1971), pp. 1–2.
2. Ibid.
3. Robert Healy and John Rosenberg, *Land Use and the States,* 2nd ed. (Baltimore: Johns Hopkins University Press, 1979), p. 2. Most states give this land-use power to local governments by means of standard zoning enabling acts. J. Douglas Peters, "Durham, New Hampshire: A Victory for Home Rule?" *Ecology Law Quarterly* 5 (1975): 64.
4. Peters, ibid., pp. 63–64. The Supreme Court upheld local zoning in 1926 in *Village of Euclid v. Ambler Realty Co.*
5. Ibid., p. 64.

[g]The coercive element of this problem may become somewhat more acceptable if the decisionmaking process allows full local participation (the aim of the balanced siting process set forth in Chapter 4), and if attempts are made to reduce these costs to a minimum and to spread them out as much as possible. (This is the aim of the compensatory approach described in Chapter 5.)

[h]A useful definition of "values" that corresponds to its usage here is given by the Organization for Economic Co-operation and Development:

> Values express underlying preferences. . . . They exist on several interdependent planes, expressing at the deeper levels a more or less conscious conception of the existential significance of human life and of relations between the individual and society, and emerging more superficially in tradeoffs between dimensions such as power, prestige, security, escape, desire for action, and so on.[111]

6. Ibid. For a treatise on the use of this power to exclude others, especially the poor, see Michael Danielson, *The Politics of Exclusion* (New York: Columbia University Press, 1976).
7. Nicholas Henry, *Governing at the Grassroots: State Law and Politics* (Englewood Cliffs, N.J.: Prentice-Hall, 1980), p. 181.
8. Bosselman and Callies, *The Quiet Revolution in Land Use Control,* p. 3.
9. Ibid., pp. i–ii.
10. Ibid., p. 1.
11. Frederick Jackson Turner, *The Frontier in American History* (New York: H. Holt & Co., 1920).
12. Silvio Lacetti, ed., *The Outlook on New Jersey* (Union City, N.J.: William H. Wise & Co., 1979), p. 391.
13. William H. Whyte, *The Last Landscape* (Garden City, N.Y.: Doubleday, 1968). Shortages of energy and water during the 1970s have intensified these changing attitudes. See William Ophuls, *Ecology and the Politics of Scarcity* (San Francisco: W.H. Freeman and Company, 1977); and Rufus Miles, *The End of the American Dream: The Social and Political Limits to Growth* (London: M. Boyars, 1977).
14. Bosselman and Callies, *The Quiet Revolution in Land Use Control,* pp. 21–25.
15. J. Gordon Arbuckle, "The Deepwater Port Act and Energy Facilities Siting: Hopeful Solution or Another Part of the Problem," *Natural Resources Lawyer* 9 (1976): 11.
16. New Jersey Coastal Area Facility Review Act (CAFRA), Chapter 185, Laws of 1973, N.J.S.A. 13: 19–1 et seq. Also see David Kinsey, "Organizing a Public Participation Program: Lessons Learned From Development of New Jersey's Coastal Management Program," *Coastal Zone Management Journal* (1980): 85–101.
17. Melvin Levin, Jerome Rose, and Joseph Slavet, *New Approaches to State Land-Use Policies* (Lexington, Mass.: Lexington Books, 1974), pp. 68–72.
18. U.S. Environmental Protection Agency, Office of Water and Waste Management, *Siting of Hazardous Waste Management Facilities and Public Opposition,* SW-809 (Washington, D.C.: U.S. EPA, November 1979), p. 144. This case is also described and discussed in Michael O'Hare, Debra Sanderson, and Lawrence Bacow, *Facility Siting* (Van Nostrand-Reinhold, forthcoming).
19. U.S. EPA, *Siting of Hazardous Waste Management Facilities and Public Opposition,* p. 187.
20. Levin, Rose, and Slavet, *New Approaches to State Land-Use Policies,* p. 65.
21. Peters, "Durham, New Hampshire," pp. 63–64.

22. Healy and Rosenberg, *Land Use and the States,* p. 6.
23. David Deal, "The Durham Controversy: Energy Facility Siting and the Land Use Planning and Control Process," *Natural Resources Lawyer* 8 (1975): 446.
24. This concept was a key component of the model code devised by the American Law Institute: *A Model Land Development Code: Proposed Official Draft—Complete Text and Commentary* (Philadelphia: ALI, 15 April 1975).
25. Healy and Rosenberg, *Land Use and the States,* p. 183.
26. Ibid., p. 212.
27. Levin, Rose, and Slavet, *New Approaches to State Land-Use Policies,* p. 70.
28. George Kush, presentation to New Jersey League of Women Voters Statewide Forum on Hazardous Wastes, Montclair, N.J., November 19, 1980.
29. On the issue of remote siting of nuclear power plants, see Marc Messing, H. Paul Friesema, and David Morell, *Centralized Power: The Politics of Scale in Electricity Generation* (Cambridge, Mass.: Oelgeschlager, Gunn & Hain, 1979) and Peter Meier, David Morell, and Philip Palmedo, "Political Implications of Clustered Nuclear Siting," *Energy Systems and Policy* 3 (1979): 17–36.
30. Richard Musgrave and Peggy Musgrave, *Public Finance in Theory and Practice,* 2nd ed. (New York: McGraw-Hill, 1976), p. 59.
31. Lawrence Slaski, "Facility Siting and Locational Conflict Resolution," *Coastal Zone '78,* vol. 1 (New York: American Society of Civil Engineers, 1978), p. 18.
32. U.S. EPA, *Siting of Hazardous Waste Management Facilities and Public Opposition,* p. iv.
33. This survey showed that the public is apprehensive regarding facilities that either generate or dispose of hazardous waste; the public also lacks general knowledge about the role of state and federal agencies in the management of hazardous substances. More than 50 percent of the respondents to the survey said they would move to another location if a disposal facility were located in their area. The respondents were also split fairly evenly on the question of whether safe hazardous waste disposal is possible at all in North Carolina. Furthermore, only 25 percent of the respondents showed any awareness of a federal role in the management of hazardous substances, and only 30 percent were aware of laws controlling hazardous substances. Nevertheless, more than 53 percent said they believed that the federal government's management of hazardous substances has been "poor" or "very poor."

    More than 56 percent of the respondents said they would rather have a single central disposal facility than five smaller ones. The poll also found that respondents who had greater knowledge of hazardous waste issues were more willing to accept wastes from distant locations at sites in their area.

The survey found a "significant" public willingness to pay for proper management of hazardous substances. Sixty-eight percent of the respondents favored internalizing waste disposal costs in product prices. More than 75 percent said the state should provide incentives for the acceptance of disposal facilities by compensating residents near such facilities for loss of property value and by granting them medical assistance. Seventy-one percent said the state should spend more of its tax funds on hazardous substances problems, and nearly 90 percent favored stiffer penalties for the illegal disposal of hazardous wastes. *Hazardous Materials Intelligence Report* (24 July, 1981), p. 3.

34. U.S. EPA, *Siting of Hazardous Waste Management Facilities and Public Opposition,* pp. 102, 108.
35. Ibid., pp. 183–187.
36. Grace Singer, "People and Petrochemicals: Siting Controversies on the Urban Waterfront," in David Morell and Grace Singer, eds., *Refining the Waterfront: Alternative Energy Facility Policies for Urban Coastal Areas* (Cambridge, Mass.: Oelgeschlager, Gunn & Hain, 1980), p. 58.
37. National Governors Association, Energy and Natural Resources Program, *Siting Hazardous Waste Facilities,* Final report of the National Governors Association Subcommittee on the Environment (Washington, D.C.: National Governors Association, March 1981), p. 2.
38. Keystone Center, *Siting Non-Radioactive Hazardous Waste Management Facilities: An Overview,* Final report of the First Keystone Workshop on Managing Non-Radioactive Hazardous Wastes (Keystone, Colo.: The Keystone Center, September 1980), p. 23.
39. Urban Systems Research & Engineering, Inc., *Handbook for the States on the Use of Compensation and Incentives in the Siting of Hazardous Waste Management Facilities,* prepared for U.S. EPA (Cambridge, Mass.: Urban Systems Research & Engineering, Inc., 30 September 1980), p. 3.

    This discussion is also found in the revised version of this document: Robert McMahon, Cindy Ernst, Ray Miyares, and Curtis Haymore, *Using Compensation and Incentives When Siting Hazardous Waste Management Facilities—A Handbook* (Washington, D.C.: U.S. EPA, 1982), p. 2.
40. Personal communication from Michael O'Hare to David Morell, 19 January 1982.
41. Booz, Allen & Hamilton, *Hazardous Waste Management Capacity Development in the State of New Jersey,* prepared for the State of

New Jersey and the Delaware River Basin Commission (Bethesda, Md.: 15 April 1980), pp. IV-24–IV-25.

42. U.S. EPA, *Siting of Hazardous Waste Management Facilities and Public Opposition,* pp. 181–188.
43. Ibid., p. 169.
44. Ibid., pp. 296–300.
45. Concern for the integrity of rural land was a particularly important factor in the decisions of Minnesota farmers to oppose a proposed electricity transmission line. See Barry Casper and Paul David Wellstone, *Powerline: The First Battle of America's Energy War* (Amherst: University of Massachusetts Press, 1981).
46. U.S. EPA, *Siting of Hazardous Waste Management Facilities and Public Opposition,* p. 204.
47. Ibid., p. 49.
48. Ibid., p. 141.
49. *Hazardous Materials Intelligence Report* (1 May 1981), p. 3.
50. U.S. EPA, *Siting of Hazardous Waste Management Facilities and Public Opposition,* p. 196.
51. Ibid., p. 214.
52. Telephone interview with Larry McCoy, General Manager–Eastern Region, Stablex Corporation, 18 November 1981.
53. The information and estimates in this paragraph were provided to David Morell by Paul Abernathy, Corporate Development Department, Chemical Waste Management, Inc., telephone discussion, 3 December 1981.
54. McCoy, 18 November 1981.
55. Gordon Bishop, "Firm Prepares Plans for Incineration Ships," *Newark Star-Ledger* (25 January 1981), p. 24.
56. Madelyn Hoffman, "Ironbound Resistance to Waste Incinerator," *Re:Sources,* Winter 1981–'82, p. 7.
57. Clark-McGlennon Associates, *Negotiating to Protect Your Interests: A Handbook on Siting Acceptable Hazardous Waste Facilities in New England,* prepared for the New England Regional Commission (Boston, Mass.: November 1980), p. 14.
58. Booz, Allen & Hamilton, *Hazardous Waste Management Capacity Development in the State of New Jersey,* p. IV-26.
59. Ronald Luke, "Managing Community Acceptance of Major Industrial Projects," *Coastal Zone Management Journal* 7 (1980): 275–276.
60. David Bradford and Harold Feiveson, "Benefits and Costs, Winners and Losers," in Harold Feiveson, Frank Sinden, and Robert Socolow, eds., *Boundaries of Analysis* (Cambridge, Mass.: Ballinger, 1976), p. 156.

61. Michael Baram, *Environmental Law and the Siting of Facilities: Issues in Land Use and Coastal Zone Management* (Cambridge, Mass.: Ballinger, 1976), p. 20.
62. Ibid., p. 34.
63. Keystone Center, *Siting Non-Radioactive Hazardous Waste Management Facilities,* p. 6.
64. William Lowrance, *Of Acceptable Risk: Science and the Determination of Safety* (Los Altos, Calif.: William Kaufmann, 1976), p. 8.
65. Chauncey Starr, "Benefit-Cost Considerations in National Planning," in Gabor Strasser and Eugene Simons, eds., *Science and Technology Policies: Yesterday, Today and Tomorrow* (Cambridge, Mass.: Ballinger, 1973), p. 247.
66. O'Hare, 19 January 1982.
67. Starr, "Benefit-Cost Considerations in National Planning," pp. 241–247.
68. R.D. Smalley, "Risk Assessment: An Introduction and Critique," *Coastal Zone Management Journal* 7 (1980): 140.
69. Lowrance, *Of Acceptable Risk,* p. 70.
70. See, for example, J.F. Byrd, "An Industrial Approach to Siting of Hazardous Waste Disposal Facilities," Speech to the National Conference on Management of Uncontrolled Hazardous Waste Sites (Washington, D.C., 15–17 October 1980); and Robert Pojasek, "Developing Solutions to Hazardous Waste Problems," *Environmental Science and Technology* 14 (August 1980): 925.
71. David Standley and Anthony Cortese, "NERCOM Issue Paper: Site Selection Process," reprinted in U.S. EPA, *Siting of Hazardous Waste Management Facilities and Public Opposition,* p. 347.
72. Keystone Center, *Siting Non-Radioactive Hazardous Waste Management Facilities,* p. 5.
73. Alvin Weinberg, "Science and Trans-Science," *Minerva* 10 (1972): 211.
74. O'Hare, 19 January 1982.
75. Owen Olpin, "Policing Toxic Chemicals," *Utah Law Review* 1 (1976): 94.
76. EPA's view has seemed to correspond to this "technological optimism." See, for example, Chemical Manufacturers Association, "Blum Warns Against 'Chemical Anxiety,'" *Chemecology* (August 1980): 11.
77. Lowrance, *Of Acceptable Risk,* p. 81.
78. State of California, Hearing: "On The Matter of Solving the Hazardous Waste Problem: Non-Toxic Solutions for the 1980's" (Los Angeles, 17 November 1980), p. 150.
79. Ibid., p. 174.

80. Clark-McGlennon Associates, *Negotiating to Protect Your Interests,* p. 10.
81. Urban Systems Research & Engineering, *A Handbook for the States on the Siting of Hazardous Waste Management Facilities,* draft, p. 40.
    In the revision of this Urban Systems report, the authors recognize that local opposition is *not* irrational. Thus, they state:

    > Even when public concern centers on identifiable potential effects of facilities, the community often views these effects differently from "experts." The public emphasizes the uncertainty of the risks and questions the ability of government, industry officials, or anyone to insure long-term safety.

    McMahon, Ernst, Miyares, and Haymore, *Using Compensation and Incentives When Siting Hazardous Waste Management Facilities—A Handbook,* p. 1.
82. In fact, there is strong evidence to suggest that even those "secure" landfills which began operation quite recently are already posing serious dangers to public health through leakage and groundwater contamination. See Peter Montague, *Four Secure Landfills in New Jersey—A Study of the State of the Art in Shallow Burial Waste Disposal Technology* (Princeton University, Department of Chemical Engineering and Center for Energy and Environmental Studies, draft, February 1982).
83. Lowrance, for one, defines risk as a matter of empirical fact. Lowrance, *Of Acceptable Risk,* p. 95.
84. See Chemical Manufacturers Association, "Accurate Communication Vital for Public Policy Department," *Chemecology* (January 1981): 6–7.
85. Keystone Center, *Siting Non-Radioactive Hazardous Waste Management Facilities,* pp. 20–21.
86. New Jersey Hazardous Waste Advisory Commission, *Report of the Hazardous Waste Advisory Commission to Governor Brendan Byrne* (Trenton, N.J.: State of New Jersey, 1980), p. 9.
87. John Hanley, "More than Superfund," *New York Times* (March 14, 1981): 23.
88. Weinberg, "Science and Trans-Science," 209.
89. Smalley, "Risk Assessment," p. 159.
90. Shelly Guyer, "The Political Pathology of Illegal Dumping of Hazardous Chemicals" (Senior Thesis, Department of Politics, Princeton University, 1982).
91. Senator Frank Dodd, Remarks during New Jersey Senate, Legislative Session, Debate and vote on S.1300, January 26, 1981. This assumption of illegal disposal versus siting policy dominates the rationale for siting new facilities; for example, Clark-McGlennon, *Negotiating to Protect Your Interests,* p. 5.

92. William Ahern, "California Meets the LNG Terminal," *Coastal Zone Management Journal* 7 (1980): 219.
93. Michael O'Hare, "'Not on *My* Block You Don't': Facility Siting and the Strategic Importance of Compensation," *Public Policy* 25 (Fall 1977): 419.
94. Lester Thurow, *The Zero-Sum Society: Distribution and the Possibilities for Economic Change* (New York: Basic Books, 1980), pp. 11–12.
95. Lawrence Bacow, *Mitigation, Compensation, Incentives and Preemption,* prepared for the National Governors Association (10 November 1980), p. 1.
96. Edmund Burke, *A Participatory Approach to Urban Planning* (New York: Human Sciences Press, 1979), p. 28.
97. R. Kenneth Godwin and W. Bruce Shepard, "State Land Use Policies: Winners and Losers," *Environmental Law* 5 (Spring 1975): 706.
98. O'Hare, "'Not on *My* Block You Don't,'" p. 419; he cites Mancur Olson, *The Logic of Collective Action* (Cambridge, Mass.: Harvard University Press, 1971).
99. U.S. EPA, *Hazardous Waste Facility Siting: A Critical Problem,* SW-865 (Washington, D.C.: U.S. Environmental Protection Agency, July 1980), p. 4.
100. O'Hare, "'Not on *My* Block You Don't,'" p. 418.
101. Thurow, *The Zero-Sum Society,* p. 213.
102. Irvin Bupp, "The French Nuclear Harvest: Abundant Energy or Bitter Fruit?" *Technology Review* 83 (November/December 1980): 34–35.
103. Levin, Rose, and Slavet, *New Approaches to State Land-Use Policies,* p. 78.
104. Danielson, *The Politics of Exclusion.*
105. Cranbury Township Mayor Scott, at a Middlesex County meeting on toxic and hazardous waste issues, 5 August 1980 (Appendix to statement by Douglas Powell, Middlesex County Planning Director, to the N.J. Senate Energy and Environment Committee meeting on S.1300, 11 August 1980).
106. Frank Michelman, "Property, Utility and Fairness: Comments on the Ethical Foundations of 'Just Compensation' Law," *Harvard Law Review* 80 (April 1967): 1166.
107. *Hazardous Materials Intelligence Report* (24 July 1981), p. 3.
108. U.S. EPA, *Siting of Hazardous Waste Management Facilities and Public Opposition,* pp. 212–213.
109. Ibid., p. 212.
110. David Morell, "Testimony on Senate Bill No. 1179" (Testimony to the Senate Energy and Environment Committee, Trenton, N.J., 28 July 1978). Also see Meier, Morell, and Palmedo, "Political Implications of Clustered Nuclear Siting."

111. William Murray and Carl Seneker, "Industrial Siting: Allocating the Burden of Pollution," *Hastings Law Journal* 30 (November 1978): 302.
112. Garrett Hardin, "The Tragedy of the Commons," *Science* 162 (13 December 1968): 1243–48.
113. Organization for Economic Co-operation and Development, *Facing the Future: Mastering The Probable and Managing the Unpredictable* (Paris, France: OECD, 1979).
114. Bradford and Feiveson, "Benefits and Costs, Winners and Losers," p. 157.

4 WHO DECIDES? GOVERNMENT AUTHORITY AND PUBLIC PARTICIPATION

Problems of political authority, of perceptions of equity and legitimacy, and of citizen participation in government decisionmaking are crucial elements in the hazardous waste crisis. The issue of siting hazardous waste facilities is fundamentally political. Conflicts over goals, values, and power are embedded in the basic tensions between majority rule and minority rights. While new hazardous waste management facilities undoubtedly will be needed in the United States in the years ahead, it is highly rational for communities and their residents to oppose waste facilities proposed for their immediate locale, and to oppose them as strenuously as possible. This inherent conflict between the majority's need for sites and facilities and the minority's right to resist being exploited in this manner lies at the heart of the politics of the siting dilemma.

One valuable way to examine the roles of the various participants in siting disputes is to focus on three principal dimensions: vertical, horizontal, and structural. Reference to these three different dimensions allows analysis of the various interests associated with any siting process, including siting of hazardous waste management facilities.

The vertical dimension distinguishes those geographical and political jurisdictions typically involved in siting: federal, state, regional, and local (county and municipal). The concept of federalism, for example, draws a vertical distinction between a state's powers and those of the "higher" federal government. This bipolar notion of federal and state political authority has certain similarities within each state, between the state government and "lower" county and municipal governments.[a] Disputes occurring along vertical lines frequently involve differences of view over the preferred level of control for particular land-use decisions, such as siting a hazardous waste management facility. Especially important is the distance between the locus of decisionmaking and the lowest level of government encompassing the actual site being considered.

The horizontal dimension differentiates between various political actors at each vertical level. Thus at times this dimension involves political jurisdictions: neighboring towns or counties, interstate disputes, and so on. Horizontal distinctions are also important, however, within single political jurisdictions, in which one finds a multiplicity of parties. In any siting proposal, for example, the concerns of the majority of a community's residents may well differ from the concerns of those residents immediately adjacent to the proposed facility ("abutters") who would be exposed most directly to whatever negative impacts (groundwater contamination, for example) the facility produced. Robert Healy and John Rosenberg have described divergent economic and social classifications at the local level in land-use disputes:

> . . . land-use controls can affect the economic prospects of different social subgroups. Intergroup conflict may arise between those whose livelihood depends on local growth and those who are indifferent to growth or actually harmed by it.[1]

The third dimension, structural, makes explicit use of these distinctions between groups. Structural divisions, or what planner Edmund Burke has deemed "functional communities,"[2] are based upon a sharing of common interests. Structural groups are relatively cohesive social or political units. Members of the Sierra Club thus are part of a structural entity; environmentalists as a whole are not. Naturally, this distinction is one of degree. Environmentalists have

[a]Under the American constitution, the several states retain a great deal of legal authority. Historically, the states joined together to form the federal union, specifically enumerating those powers to be provided to the central government while reserving all other powers to themselves. In contrast, no state was formed in this manner by a grouping of local communities. Instead, state governments delegated certain powers to local governments, primarily by statute although occasionally in state constitutions—this varies from one state to the next. Such statutory delegations of authority can be revised at any time by passing a new statute. Despite these historical and constitutional differences, many significant parallels exist in practice between federal-state and state-local relations.

greater structural unity than does a town's entire population. Analyses of decisionmaking often refer to structural distinctions concerning groups involved in the outcome: politicians, interest groups, the public-at-large, for example. Any individual, of course, may belong to more than one structural group.[3]

THE CAST OF CHARACTERS

According to Michael O'Hare, "facility-siting debates vary in detail but they usually conform to a standard model with a predictable cast of characters."[4] A simplified level of analysis suggests that the following interests dominate the siting process for hazardous waste facilities: (1) state agencies charged with siting responsibility; (2) the facility developer, typically a firm in the waste management industry; (3) the prospective users of the facility (the chemical industry and other waste generators); (4) the community in which the site is proposed; and (5) statewide environmental and public interest groups. Figure 4–1 depicts these five major groups, who all played an essential role in formulating new siting legislation in New Jersey: the chemical industry, the hazardous waste industry, environmental and public interest groups, local officials, and the state's environmental agency.

Discussions of hazardous waste facility siting, however, too often ignore the horizontal composition of these groups.[b] It is true that in many communities that possess potential sites for hazardous waste facilities, the proposed facility presents such a "bad bargain" that nearly everyone in fact does oppose its development; only one set of interests emanates from community residents. Nevertheless, the subtleties of potentially divergent intracommunity interests must be recognized in designing mechanisms for more representative local participation in siting decisions. Lack of attention to these horizontal differences within particular communities may cause serious difficulties when siting is actually attempted.

A more complete set of the different interest groups involved in siting, therefore, includes the following:

- developer;
- facility contractors;
- labor;

[b]Particularly problematical is the fact that community interests are assumed to be homogeneous.[5]

Figure 4-1. Groups Involved in Facility Siting: Three Dimensions (Simplified).

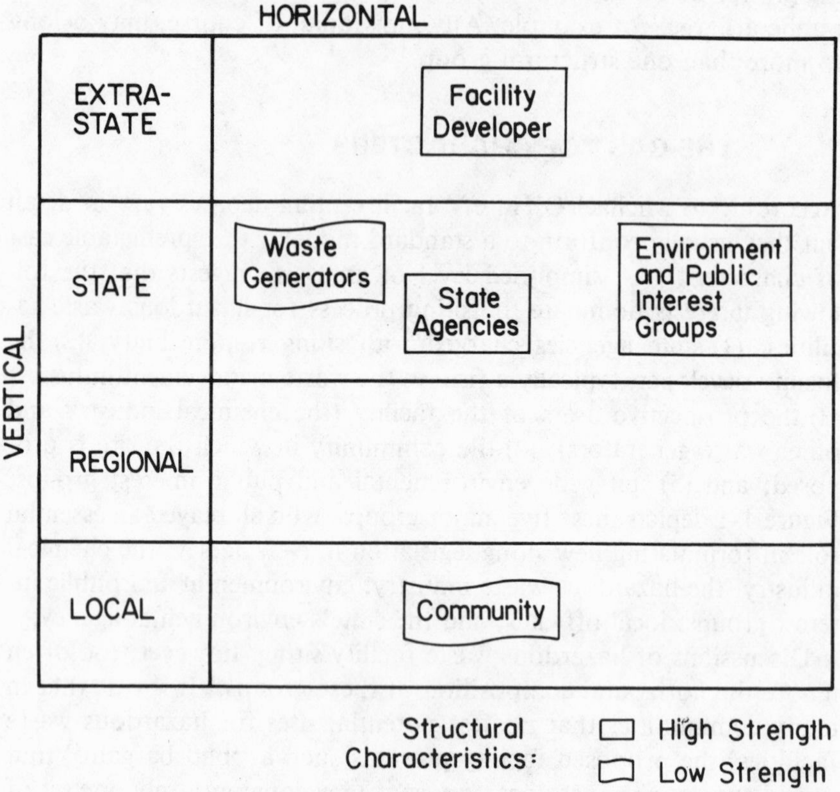

VERTICAL / HORIZONTAL	
EXTRA-STATE	Facility Developer
STATE	Waste Generators; State Agencies; Environment and Public Interest Groups
REGIONAL	
LOCAL	Community

Structural Characteristics: ▭ High Strength; ⏥ Low Strength

- industrial hazardous waste generators;
- environmental and public interest groups;
- state politicians and agencies;
- regional or statewide support (those benefiting from "safe" disposal or from industrial expansion);
- regional opposition;
- local politicians and planners;
- local support (those favoring lower property taxes, potential community contractors or service providers);
- local opposition (those whose property would be appropriated for sites—"takees," abutters, others who fear adverse impacts).[6]

Figure 4–2. Groups Involved in Facility Siting (Differentiated).

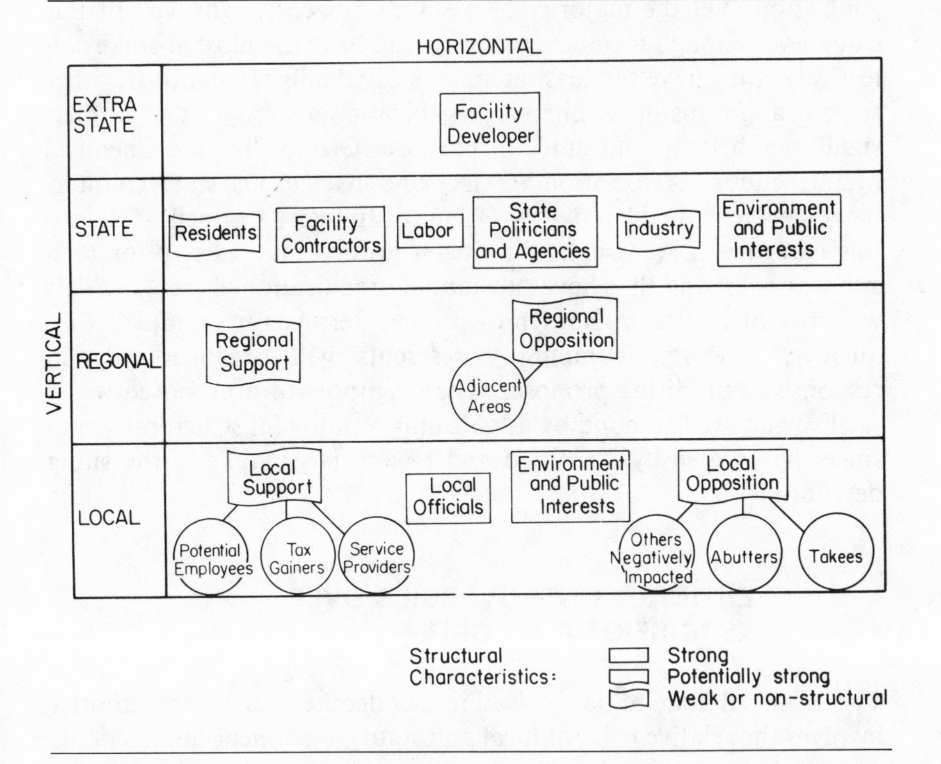

Note: Sizes of boxes are not meant to indicate relative size of groups. Were this done, the box at the state level showing "residents" would be many times the size of the rest. The developer in this diagram is shown to be from out of state. This needn't be the case (could be shown at state level). In New Jersey, however, both SCA Services and Rollins Environmental Services are corporations which have their national offices outside the state.

These eleven types of participants are depicted in Figure 4–2. The configuration of several potentially distinct interests at the local level stands in sharp contrast to the single "community" grouping shown in Figure 4–1.[c]

"Statewide support" for a new hazardous waste facility is undoubtedly the weakest of all these categories in structural terms. This

[c]Even more detail on horizontal interests could, of course, be provided—the extreme being a person-by-person breakdown. This would be too detailed and cumbersome to be worthwhile, however. The other extreme of one homogeneous "community interest" often employed is too facile, though.

grouping comprises a very diffuse constituency that lacks any real organization. Yet the majority of a state's citizens might well fit this category.[7] Though members of this group have the most at stake collectively, they have the least at stake individually. In contrast, other structural groups have inherent organizational strength due to their small membership and uniform purpose. Groups like the Chemical Manufacturers Association (CMA) and its various state counterparts—such as the New Jersey Chemical Industry Council—fall into this category. Likewise, some citizen groups may exist prior to a siting attempt and thus have substantial organizational strength. This was true of HOPE in Bordentown, New Jersey, for example.[8] Frequently, however, community residents will organize only in response to the siting proposal. The neighbors of a proposed waste facility may well respond by establishing a powerful structural group since they are easily identified and have a large stake in the siting decision.[9]

VERTICAL CONTROVERSIES OVER EXERCISE OF AUTHORITY

The principal issue to be resolved in any debate over siting authority involves the relative roles of local and state government jurisdictions. This decision is critical to the development of a state siting process which can gain the respect of all the parties involved. Should the local government hold veto power over the siting proposal? If not, should the state exercise the power to *override* local rejection of the proposed facility, or even the power to *preempt* local authority entirely by making the yes or no decision at the state level? And if a powerful role for the state is envisioned, perhaps because of a perceived need to ensure that an adequate number of sites and facilities are indeed available in the face of local opposition, what should be the composition of the state decisionmaking body—and what processes should it establish and follow?

As noted earlier, control over land-use decisions in the United States has been delegated quite broadly to local communities through state standard zoning enabling acts and similar state legislation. The lack of success in recent attempts to site hazardous waste facilities can be attributed, in large measure, to the dominance of local control over these proposed land uses. Naturally, such local "home rule" power militates against approval of sites whenever they threaten local interests. Lack of approval from the local zoning commission of Baltimore County, as one example, kept Allied Chemical Company from locating its proposed new landfill in Rossville, Maryland.[10] Similarly, the three-to-two decision by the Hooksett, New Hampshire, Planning Board to reject a Stablex Corporation waste solidification facility has been a principal obstacle to this company's siting proposal.[11]

The goal of most siting reforms is to devise a means to bypass such local vetoes. As perhaps the latest trend in the "quiet revolution" in land-use control, a number of states have begun to assert their right to shift decisionmaking on new hazardous waste sites away from local actors to the hands of state agencies. In theory, this higher level of decisionmaking can better take account of overall statewide interests. Fred Bosselman and David Callies noted that local zoning is "woefully inadequate to combat a host of [land-use] problems of statewide importance."[12] This view provides the logical antecedent to EPA's recommendation that "the process of site selection should not be hampered by blanket local vetoes."[13] The Chemical Manufacturers Association (CMA) likewise feels that inadequate consideration of state interests has been primarily responsible for the fact that insufficient sites are available for new hazardous waste management facilities:

> In most states local land-use regulatory authorities, cities and counties, can veto the location of a new facility by withholding the necessary discretionary zoning or other land-use approval. Although one court has held that RCRA and qualifying legislation preempt local regulation, many states have sanctioned a double veto system which requires both state and local approval before a disposal facility can be sited. This double veto usually precludes siting new facilities because local authorities would not balance local desires versus statewide interest in both environmental quality and industrial development. Thus, there is a need for state intervention in the facility-siting process to provide a statewide forum for the resolution of local industry conflicts and to balance the general public interests in environmental quality and industrial development.[14]

Localities are thus perceived by many observers as wielding too much power in decisions over sites. Hence, there has been growing pressure across the country to shift the balance of siting power away from the local level to the state in order to allow for establishment of needed facilities. In J. Gordon Arbuckle's view, ". . . it may sometimes be necessary to put a legislative thumb on the scale in order to permit timely construction of truly essential projects."[15] The most obvious means for a state to put its "thumb on the scale" and shift the vertical dimension of siting authority upward is to establish a process whereby local control can be avoided, either by state override or even by state preemption when the state's interest is jeopardized.

The distinction between *override* and *preemption* is critical to this analysis. Preemption involves placing all decisionmaking authority in state hands. While local residents and county and municipal government leaders can participate in the preemptive state decision, they have no power to reach their own independent decision on the siting proposal. In contrast, override involves sequential decisionmaking. Under this arrangement, the local government can assess the particular siting proposal in its own area and decide whether to approve or reject it. The state then has the authority to review this local decision and either affirm it, approve it with modifications, or override it, in the interest of the state as a whole.

This override process has a parallel in typical relationships between the legislative and executive branches of American government. The legislature passes a bill or resolution (akin to the initial siting proposal); the executive may veto it (analogous to a local government's rejection of a proposed facility); and the legislature may override the veto (as in state override of a local rejection). While the local government's role in an override process is certainly not autonomous nor absolute, it is far greater than under a preemptive arrangement.

State Siting Laws and the Myth of Preemption

William Murray and Carl Seneker advocate creation of a "master siting agency" to shift land-use authority to the state level, a recommendation that directly reflects the American Bar Association's endorsement of "super siting agencies."[16] The preemptive aspect of these agencies is derived from their legislatively established power to ignore local zoning and impose needed sites for facilities within selected communities, usually by means of the state's eminent domain powers.

Several states have adopted legislation to override local control over the siting of hazardous waste facilities. The original draft of New Jersey's hazardous waste siting legislation (S.1300) exemplified this preemptive approach, providing for a new "hazardous waste facilities corporation" to hold condemnation powers to acquire land needed for a facility. Explicit preemptive language allowed land to be obtained "without regard to local zoning ordinance."[17] Introduction of this legislation followed reports by the Governor's Hazardous Waste Advisory Commission and the management consulting firm of Booz, Allen & Hamilton, which both recommended that explicit eminent domain powers (along with certain safeguards on their use) be provided to a state agency—even though the state *already* had limited preemptive authority to site hazardous waste facilities.[d] Michigan's Act 64 established a Site Approval Board empowered to preempt local ordinances and decide whether to approve or reject permit applications for hazardous waste facilities. This preemptive power is tempered only by the statutory directive that the board consider whether facility proposals offer "consistency with local planning and existing development."[19] This decision was a direct result of Michigan's tragic experience with toxic pollution from polybromated biphenyls (PBBs), and the political sensitivity which followed. As Governor William Milliken stated:

> We have learned from the PBB experience that we need to control the use of chemicals for our own benefit. The legislation that has been enacted in the past three to four years—establishing the Toxic Substances Control Commission and coordinating responses to future chemical problems—is far ahead of other states. We still have much to learn, but we have established the basic system for dealing with toxic chemicals. . . .
>
> We have learned through our experience with PBB that we cannot take chemicals for granted. Chemicals are an integral part of our lives, and contribute greatly to bettering our quality of life. But we recognize now that we have to learn how to control the possible ill effects of those chemicals that we encounter daily.
>
> Michigan has put into practice a system of control of toxic chemicals that is a leader among other states. The Toxic Substance Control Commission, at a recent conference on toxic waste, was considered to be a model for other states.

[d]The state Solid Waste Authority could approve a developer's proposal for a site in spite of local rejection of the facility. The state could not actually obtain the land for a facility, however.[18] (See Chapter 2.) This resulted in a lack of positive siting authorty for New Jersey's Department of Environmental Protection (DEP), and in pressure for a more preemptive state statute.

> Our lifestyle generates wastes, and it should be a concern of everyone—not just government—that we take an active role in dealing with chemicals and their possible effects. We cannot have the mentality of "pushing off" the problem to some other city or state.
>
> Even though we are committed to improving the business climate of our state, our first priority must be the protection of our people from toxic chemicals. The PBB accident caused a massive amount of physical and economic hardship to thousands of people, and we cannot afford inaction or passivity in approaching the need for sensible handling of toxic chemicals.[20]

Several other states have passed legislation creating preemptive state siting agencies or otherwise allowing state government agencies to bypass local ordinances in order to obtain sites. Among these states are Arizona, Connecticut, Florida, Kansas, Maryland, Minnesota, New York, Ohio, Pennsylvania, and Tennessee.[21] All of these statutes are recent, having gone into effect since 1980.[22]

A master siting agency and preemptive state power are attractive in that this approach seemingly guarantees a means of obtaining needed sites, since it removes the local zoning obstacle.[23] An essential question, though, is whether this preemptive strategy actually will prove successful in obtaining sites. Since no state has yet obtained a site by means of a preemptive process, judgments as to its effectiveness remain largely speculative.[24] Yet analysis of siting controversies suggests that this state power may well prove illusory in the face of determined local opposition. The problems encountered in Minnesota have already been noted.[25] Lawrence Bacow has described how the Massachusetts experience exemplified the realistic constraints associated with preemptive measures:

> Massachusetts flirted briefly with the preemption approach to hazardous waste facility siting. While the legislature debated a preemption bill, the state commissioned a consultant to identify the most feasible and desirable sites within the state for a hazardous waste disposal facility. Within days after the consultant filed his report, legislators from the three cities recommended by the consultant filed legislation to statutorily exempt their cities from further consideration. The bills passed both houses and were signed into law by the Governor. The lesson from this story is that *while the state can strip local officials of their permitting power, it cannot strip them of their political power which can also be used to defeat an unwanted facility.*[26]

Continuing difficulties in siting are to be expected. They will only be intensified when states rely on a preemptive mode of hazardous waste facility siting. In fact, the notion of a state being able to guarantee the availability of sites regardless of the degree of local opposition seems wholly unrealistic. As the first Keystone workshop participants concluded: "Many students of the problem believe that in the current political milieu, the preemption of local authorities often will be ineffective or counterproductive."[27]

It seems safe to assume that if a locality is vehemently opposed to a proposed hazardous waste facility, the project will *not* be successful, regardless of the existence of state preemptive (or override) powers. One reason lies in the potentially effective use of legal measures to oppose unwanted facilities. Alan Farkas has warned that:

> . . . although state preemption may remove one mechanism to stop a proposed facility, citizens may be able to find other mechanisms such as the National Environmental Policy Act of 1969 or similar state requirements which can be used to delay or even stop a proposed development.[28]

Judicial review of state siting actions undoubtedly will be occasioned whenever local citizens oppose a state's attempt to site a hazardous waste facility in their community. Indeed, New Jersey's siting statute, as rewritten, explicitly recognizes that the decisions of the state siting commission will be subject to judicial review.[29] Judicial intervention in siting decisions, however, is frequently the "death knell" for these projects.[30] Lester Thurow described why legally imposed delays frequently have secured a project's demise:

> To be able to delay a program is often to be able to kill it. Legal and administrative costs rise, but the delays and uncertainties are even more important. When the costs of delays and uncertainties are added into their calculations, both government and private industry often find that it pays to cancel projects that would otherwise be profitable.[31]

Another important class of tactics used on occasion to oppose hazardous waste facilities might be termed "quasi-legal." After protracted battles, a facility may well be sited in a locality by state override or preemptive measures. Nevertheless, the facility owner's problems might be far from over, since an unreceptive community could cause continual problems for the facility. In fact, it may be disadvantageous for a facility developer to avail himself of state condemnation powers to obtain a site unless local residents and the host community are willing to acquiesce, at least to some degree, to the operation of the facility.[32]

The case of an Earthline, Inc., landfill in Wilsonville, Illinois, provides a rather unusual illustration of this proposition. After having been in operation for several months, this facility encountered large-scale public opposition to its plan to dispose of PCB-contaminated soil from Missouri. While carrying on legal proceedings against Earthline, the town managed effectively to halt the facility's operations by digging a trench across its access road. This action—a quasi-legal measure—was carried out, officially, as one part of culvert repair work needed to control flooding.[33]

Local opponents may employ a third set of tactics to oppose attempted siting of hazardous waste facilities. These "extra-legal" actions might include sabotage, violence, and other illegal activities on the part of those desperately opposed to the prospect of living near hazardous wastes. Bacow concludes that

> If people are fervently opposed to a hazardous waste facility in their town, they will find some mechanism for making their feelings known. It is not inconceivable that the civil disobedience tactics that have been employed against nuclear power could also be employed against hazardous waste facilities if the state tries to force localities to accept them. Passions run high when people feel their health is threatened.[34]

In several cases, those who perceive themselves as actual or potential victims of dangerous toxic chemical substances have resorted to this extreme class of tactics. Events in Wilsonville, Illinois, showed that violent public resistance to facility operation is a distinct possibility. At one point, Wilsonville residents gathered in an angry mob outside the facility where "some in the crowd reportedly . . . threatened to blow up the facility."[35] In Jackson Township, New Jersey, a septic tank cleaning business suspected of dumping toxic pollutants into a municipal landfill reportedly received threatening phone calls and incurred vandalism from angry residents concerned about the presence of these pollutants in their water supplies.[36] These more extreme forms of opposition should not be overlooked as possible means by which some citizens will resist state preemptive measures.

Barry Casper and Paul David Wellstone have described in some detail the growing frustration of basically conservative Minnesota farmers aggrieved over a siting process that brought an unwanted high-voltage electric line through their region. These farmers eventually resorted to violence, first by blocking construction crews from access to the area and then by shooting insulators off of the lines and by pulling down some of the power line's huge towers.[37]

Because of the considerable range of means—legal, quasi-legal, and extra-legal—available to local opponents to prevent siting of unwanted facilities, preemptive measures whose goal is to establish sites through assertion of exclusive state power are not likely to enjoy much long-term success. They establish administrative efficiency but at a very high price in terms of political equity and popular acceptance of the siting process. State policymakers would thus do well to consider the conclusion reached by Murray and Seneker about the politics of the preemptive stance to land-use decisions which they recommend:

> As a practical matter it should be noted that it almost surely would be politically impossible to impose a major industrial facility, or to propose legislation which would allow such a facility to be imposed, on an area that does not want one. Despite the statutory authority which a state may have to impose a facility on a given area, vehement local opposition will always provide a political veto to truly unpopular action.[38]

In sum, holding a permit or a legal right to proceed does not guarantee that the project will proceed to completion. Such permits are a necessary but far from sufficient prerequisite to successful siting of a new facility in the face of determined public opposition.

Legitimate State Intervention: A Balanced Siting Process

Although political realities may dictate against reliance on a purely preemptive approach to siting, it is evident that the states must play a leading role in the siting process. The alternative—a continued pattern of parochial local decisions in which facilities of statewide benefit will, all too often, continue to be rejected—is unacceptable. What balance is possible between abject preemption and total local veto power?

The overall siting objective must to be ensure that the decision-making process is perceived as legitimate by the residents of the local area. Although some of them may never fully accept the presence of the new waste facility, their respect for the siting *process* is a necessary minimum condition.[39] As one individual close to the issue of hazardous waste management in New Jersey put it in late 1980: "People are basically fair-minded. If they are treated in a fair way, they will respond accordingly."[40] More pragmatically, the political effectiveness of those who continue to oppose the proposed facility will be constrained if many people in the local area have decided to respect the validity of the siting process. In this respect, a process based on state override has a much greater chance of gaining local respect than does a process based on state preemption.

Since state constitutional dominance over land-use decisions is clear, the states certainly have a right to intervene in local land-use decisions in order to protect nonlocal interests. More critical is the *form* this intervention should take. Here the distinction between preemption and override becomes crucial. In most states, community residents and local officials are not necessarily opposed to all measures which would allow state interests to be included in land-use decisions. This was evident in New Jersey's process of shaping siting legislation, for example. In testimony before the New Jersey Senate Energy and Environment Committee, the director of Sussex County's Freeholders stated:

> No one likes condemnation, and we feel that every effort should be made by the commission to exhaust every avenue before getting involved with condemnations. Nevertheless, condemnation should be used when all else fails in order to promote the health and safety of our citizens.[41]

As Los Angeles Mayor Tom Bradley said in a California hearing on hazardous waste in late 1980:

> For one who through all of his political career has joined with other local government officials in preserving or fighting for home rule, I think this [hazardous waste management] is one point at which we're going to have to take another look. It's just a question of the realities in front of us. And I think that some kind of state guidelines are going to be required if there is any hope that we can deal with this question either on a local or regional basis.[42]

This local receptivity to the exercise of state authority in siting certainly does not amount to approval of broad-based state preemption. Rather, it is a recognition of the legitimacy of limited state override measures, combined with extensive local participation.

Local residents object to state preemption in large measure because it places priority on administrative efficiency over political sensitivity. Preemptive state mechanisms lack accountability to local interests.[43] Thus people see state centralization of decisionmaking as insensitive and "undemocratic."[44] Despite their limitations, local control and home rule are attractive because they afford maximum public access to and influence on the decisionmaking process.[45]

State intervention in siting can be effective without resort to preemptive measures. J. Douglas Peters, for one, advocates a more balanced approach:

> The natural tendency is to approach the home rule question as an all or none proposition, and it need not be so. Equity suggests that the state should not decide what is good for the town when the town has an interest, anymore than a town should decide what is good for the state when the state has an interest. Systematic forethought allows the deliberate measure and balance of state and local benefits.[46]

A few states seem to have adopted a more balanced view of state intervention, falling between local vetoes and state preemption. The principal approach is to rely on override rather than preemptive siting powers. Connecticut's HB 5400, for example, provides its state siting board with power to override local planning. Unlike Michigan's Act 64, here the facility applicant must first undergo scrutiny by the local zoning board. Only after receiving a local veto can a developer request the state to exercise its override powers.[47] Florida's HB 311 similarly stipulates that a developer's proposal must first be submitted to local government planning agencies. If the siting proposal does not comply with local ordinances, a developer's request for a variance must then be approved by the regional planning council, after which it can go before the state's governor and cabinet for final decision.[48] Although Connecticut's state siting power resides with a siting board and Florida's lies in the executive branch, the two states both provide for state override of local vetoes rather than for state preemption.

Experience in siting controversies across the country suggests that several specific actions will contribute to the creation of a siting process which will be seen as legitimate by many of the people who live near the proposed site. Basically, the best hope lies in creating a *balanced, sequential,* and *timely process* of decisionmaking in which all of the parties concerned—both the majority and the minority—have a clear opportunity to express their legitimate concerns. Compromise and balance are needed to avoid the abuse of power by any single level of government. Complete home rule with local veto authority is unacceptably parochial; unbridled state—let alone federal—preemptive powers are unacceptable as well. The process will fail if excessive, unbalanced authority is given to either sector.

Effective communication, timeliness, and procedures for an open dialogue are all essential elements of an effective siting process. People must be convinced, by evidence presented in the siting process, that the proposed facility really *is* needed. And in the end, after all the parties have had a fair opportunity to express their opinions, authority must exist for a clear decision to be made so that the facility can be built. As the Keystone Center conferees concluded: "The entities with authority and responsibility for making decisions must be clearly identified."[49] Someone has to bear the eventual burden of making such a tough choice.

In every state, new siting legislation would be required to create such an authoritative statewide decisionmaking process. The legislation should encompass approval for waste management facilities within industrial complexes ("on-site facilities") as well as "off-site" regional disposal facilities. This siting process forms the heart of the political power structure around which revolve issues like compensation, technology and safety assurances, siting criteria, procedures for public participation, and so on.

The core of this concept is a balance of political authority. The new state legislation must establish clear authority for the local government to review any hazardous waste facility proposed in its domain—not only under local zoning powers, but more broadly with regard to overall local consent or opposition.[e] The issue has far less to do with the legal and procedural technicalities of zoning than with the basic wishes of the local governing body. Quite simply, do the elected local government and its constituents approve of the facility proposed for their area, or not? The new state legislation ought to require formal local approval (or rejection) of any proposed new hazardous waste facility even if the area of the proposed site is not zoned (including, for example, sites in the unincorporated areas of many counties).

[e]New Jersey's siting legislation, for example, provides for extensive local review and comment on the siting proposal, prior to exercise of state preemptive decisionmaking authority. Because of its broad provisions for public participation in such local review, this process may come quite close to the legitimacy goal. However, such legitimacy might be achieved with greater certainty by use of override rather than preemption (see Appendix B).

From a local perspective, of course, provision for state override powers is vastly different from continuance of local veto authority. Will people really understand the subtle but crucial distinction between override and preemption? In Massachusetts, for example, some observers believe that community opponents have reacted with equal intransigence to the apparent existence of state override authority (through mandatory arbitration) as they would to state preemption.[50] Such opponents prefer a clear local authority to reject a proposed hazardous waste facility. In the face of a need for new facilities, however, reliance on local decisionmaking generally does not work either. The closest one can come to a legitimate process which can resolve the siting dilemma seems to be to establish an override authority rather than either local veto or state preemption.

In devising new legislation, each state will face several important issues: precisely how much siting power it wants to share with its municipalities and counties; whether local authority should be exercised by municipal or county government (or both); whether an executive (mayor) or legislative (city council) decision is preferable; and what role should be played by regional government entities (councils of government). However these matters are resolved, *the fundamental principle involves the grant of authority by the state to the local community to say "yes" or "no" to the proposed hazardous waste facility.*

A local decision of this kind is vastly different from local participation in state preemptive decisions, in which the actual authority rests with a state agency, board, commission, hearing examiner, or administrative law judge. Experience across this country indicates that citizens feel isolated from state (and federal) decisions on siting proposals. These decisionmakers are far too distant from the local political arenas in which citizens are accustomed to acting. In contrast, even in the 1980s, citizens in many communities feel able to call the mayor at home or to drop by the city council member's office on their way home from work.

Given the level of emotions fed by fear of hazardous waste facilities, explicit local decisionmaking authority seems imperative. The local government must be allowed to examine the siting proposal, conduct any necessary studies, hold its own hearings, and reach its own decision. The community should be encouraged to hire its own expert consultants to advise on the siting proposal. This will enhance citizens' confidence in the validity of the process. In contrast, their suspicions will be heightened if the local government has to rely solely on information provided by the industry or a state agency. The state might provide financial assistance to the local government for this purpose; or the necessary studies might be funded by the facility sponsor through permit application fees.[51]

Clearly, local decisions on proposed new hazardous waste facilities frequently will be negative, given people's attitudes at present. As Mayor Bradley said at the 1980 hearing in Los Angeles: "That is one of the political realities. . . . No city, no local community is going to gladly or willingly accept disposal."[52]

Why would a local government *ever* say yes? One obvious inducement for local governments to give serious consideration to approving a siting proposal would be a state siting law that allowed the local government to include certain kinds of special facility operating conditions and requirements in its approval permit. If the local government chose to reject the facility outright, it would lack any formal power to insist on such conditions in a subsequent state override of the local rejection.

In this regard, the allocation of authority in a balanced siting process reinforces the effective use of negotiation between the facility developer and the host community. Given the need for new hazardous waste management facilities, and the existence of ultimate state override authority, the balanced process can encourage local decision-makers to focus on what can be packaged with the proposed facility and on what protections can accompany it. Such creative redesign of the initial siting proposal—by technical alterations, compensation measures, and so on—is vital to achieving a facility which can be acceptable all around. The balanced siting process helps accomplish this vital objective.

Expanded public participation procedures in a preemptive siting process are a far cry from such a balance of state and local authority. The revision of New Jersey's siting legislation, for example, focused on this issue, producing greatly expanded opportunities for local participation. Unfortunately, respect for the siting process may not result from such changes. Indeed, in their analysis of the Minnesota powerline dispute, Casper and Wellstone show that, ironically, new procedures established by the state to enhance public involvement in siting had the opposite result, leading to lessened public acceptance:

> Some of the farmers . . . were no longer in a mood to work within the system. Many had expended an immense amount of time and energy

> "participating" in the state proceedings, which they now felt to have been a charade. [One farmer articulated] the anger and bitterness that many farmers felt: ". . . They say they're giving the people input or whatever; they'd be a lot better off just to decide to put it over here, because we had no say-so anyway."[53]

There is no easy substitute for the balance of state and local authority over siting these controversial facilities.

Moreover, the local government needs to have an opportunity to reach its decision prior to announcement of any formal state decisions concerning the proposed site. Such a requirement constitutes the sequential aspect of the process. This will ensure that local decisionmaking is not preempted by the actions of a state agency.[54] Otherwise, local decisionmakers will feel intimidated by the state's decisions, which are based on the presumed expertise available to state environmental and health departments—and local respect for the siting process will diminish accordingly. State decisionmaking can occur after the local government has had a reasonable chance to express its concerns and to act in pursuit of its own best interests. State expertise could be made available to local decisionmakers through environmental impact assessments, risk analyses, and so on—while still allowing the local community the right to express its own wishes on the pros and cons of the siting proposal.

Starting the siting process with an unambiguous local decision provides far more opportunity for effective political activity by local citizens than does any other model of participation or representation. Many other approaches have been suggested to incorporate local concerns into a preemptive state siting decision: appointing elected officials to a state siting board, seating residents of the area on the board during its consideration of specific sites, requiring public hearings to be held in the local area, and so on.[55] All of these concepts, however, pale in comparison to allowing for a valid local decision in the siting process. While frequent state override of local rejections might diminish local respect for the siting process, shifting the decisionmaking authority totally into the hands of a state agency—even with opportunities for intense local involvement—is even worse. This step would itself automatically diminish respect for the siting process within the local areas chosen. In this sense, the difference between override and preemption lies in the formal and informal constraints on such state action, and in the way local opponents of the facility perceive the legitimacy of the overall siting process.

Timeliness is the third component of the process. A balanced siting process should not allow local vetoes through inaction. Therefore, the state siting statute probably should restrict the time available to the local government to reach its decision. In most cases, six or perhaps nine months would seem an appropriate deadline for a local decision to be reached, since these decisions may require the commissioning of studies by outside consultants. If the local government has not reached a clear decision after six or nine months, it could be deemed to have approved the proposed facility by default. Such a provision in the state siting law would provide quite an impetus for timely local action.[f]

Obviously, one must distinguish between a siting process which is accepted as legitimate by people who might at some future time be affected by it, and one which retains its legitimacy even after it has operated in the locality. The latter is the most important goal. Many people involved in discussions about siting express the belief that if you can get someone to agree to some process at the beginning, he will in some way be bound by that agreement as the process is implemented. The experience with site identification by exclusionary criteria evidences that just the opposite is the case—that people will agree to all sorts of things and then change their mind and rescind their agreement when they discover the total costs involved. The only true test of the legitimacy of a siting process, therefore, lies in its use.[56]

While a balanced siting process must incorporate a legitimate and meaningful role for the affected local governments, in the interest of the state as a whole, it must also include *authority for state override of the parochial local veto.* This override will be difficult politically for the state, to be sure; even more perhaps than if a state agency had made the initial decision itself, in a preemptive manner, without prior local action. Such "difficulty" is desirable, however. Given the political sensitivity associated with all state land-use decisions and the emotional climate associated with hazardous wastes, legislation which allows a state agency or state siting board to make this decision directly may well gain administrative expediency, but at the expense of public acceptance of the siting process. Eminent domain is no sub-

[f]The state's siting statute would have to define a consistent starting point from which to measure the time used for local review. This could vary from one state to the next. The basic intent is to give the local government ample time to review the proposed site or facility and reach a formal yes or no decision, and to do so in a timely manner.

stitute for effective political involvement by concerned citizens, at the local as well as the state level.

These informal constraints on state exercise of its override powers should be explicitly reinforced in the siting statute by a set of formal procedures designed to ensure that override is a difficult task. For example, the Congress must use a two-thirds vote rather than a simple majority to override a president's veto of legislation. Similar provisions could apply to state override of a local community's rejection of a proposed hazardous waste facility—two-thirds of the members of a siting board, for example, or of the state legislature.

States have a number of options in choosing an appropriate institutional mechanism to use in overriding a local decision. This choice will be an important component of each state legislature's debate over the hazardous waste siting bill, and a great deal of variation from one state to another is likely to emerge. The power could be assigned to the governor directly; to the head of an existing state agency, to a new siting board (composed of citizens as well as executive branch officials), to a new state-chartered hazardous waste management commission, or even retained in the hands of the legislature itself (although the inherent political pressures here may be more direct). The experience of many states over the past decade in developing procedures for decisionmaking for land development in their coastal zones may be a useful indication of the types of issues to emerge. Each approach will have advantages and disadvantages in accommodating both public participation and expertise.

Internecine warfare among competing state power blocs over the locus of this state authority should not be allowed to divert attention from the broader principle of balanced state and local authority, and of sequential decisions from the local level up. The appointment of a few citizens to a state board which holds unilateral siting authority (preemptive rather than override authority), or expanded procedures for state public hearings within the local area will contribute much less to the creation of local respect for the siting process than assignment of initial decisionmaking power to the local government.

Regardless of whether a state is authorized by EPA to administer the federal RCRA permit program or not, local or state land-use permit approvals will still be required. Therefore, every state should devise its own version of a balanced, sequential, and timely siting process, irrespective of the state's legal authority under RCRA.

The balanced state-local siting process should apply to new on-site industrial waste facilities as well as off-site regional facilities. Both kinds of waste management facilities will be needed. Public opposition can arise to on-site as well as off-site proposals (though the latter seem even more vulnerable than the former). Therefore, both deserve the same level of public scrutiny.

Creating a balanced, sequential, and timely process for reaching decisions on siting new hazardous waste facilities will not automatically resolve all of the siting problems, of course. A wide range of controversial issues will affect local attitudes toward siting, and these issues must be addressed in overall national and state policies toward hazardous waste management. The politics of siting, however, lie at the core of the dilemma. In contrast to preemptive schemes, a balanced siting approach is more likely to meet local interests. By allowing for an active local role as an integral part of the siting process, local residents may well come to regard the decisionmaking effort as legitimate. The importance of establishing a legitimate procedure should not be underestimated. As Burke has cautioned, if planning activity is to be successful, it is more important that those who are affected sanction its legitimacy than that it have the backing of authority.[57] The path to legitimate yet successful location of hazardous waste facilities appears to lie in the establishment of an undeniable state siting presence balanced against the preservation of a meaningful local role.

PURSUING THE HORIZONTAL DIMENSION

Debates over who ought to be involved in siting decisions have tended to focus on the vertical dimension, especially on the degree of authority that the states can legitimately wrest from localities to ensure that sites are available. Nevertheless, implicit in any siting process is also an allocation of decisionmaking power to various groups at each vertical level. Elaboration of these horizontal interests and their potential conflicts helps make this allocation explicit, bringing it into line with overall siting goals and decisionmaking values.

The simplest model of participation in facility-siting decisions centers on just three main parties: the developer, community officials, and state officials. These tend to be the groups with the greatest structural strength. Advocates of representative democracy are, at least in theory, likely to be basically satisfied with this simple model, since local officials presumably speak for local interests and state officials speak for state interests. Everyone who has a stake in siting is therefore represented formally in the decisionmaking process.

To those more inclined to the ideal of participatory democracy, however, the simplicity of this basic model signals its obvious deficiencies. According to this viewpoint, attempts should be made to "open up the process" in order to maximize participation from the different groups shown in Figure 4–2. An evident advantage of this approach is that the individuals and groups with an interest in siting are given a more direct part in the decisionmaking process. One major disadvantage is the potentially low level of knowledge among those who are given greater involvement in decisions. This lack is especially apparent with respect to the technical issues associated with construction and operation of hazardous waste facilities. Another important constraint to implementation of this democratic ideal is the possibility that the process will be rendered time-consuming and inefficient due to an abundance of participation. Moreover, those who participate have an incentive to do so—others with a lower level of individual concern may not participate.

Community interests in siting are typically quite varied. A report issued by the New England Regional Commission is one of the few that recognizes that community interests regarding hazardous waste facility siting are not necessarily homogeneous:

> There are . . . many different interests within the community that are going to react differently to the proposal. Some will look at it as an opportunity and some will look upon it as a disadvantage. An abutter has one point of view. If you live in that neighborhood but you are not an immediate abutter, you have another point of view. If you're the health department you have another point of view. If you are the conservation commission you have still another point of view. The list of possible positions and points of view is lengthy. It is impossible for the chief elected official to formulate a. . . .position for "the" community since "the" community is a collection of diverse interests and not a single interest.[58]

Perhaps the most important issue in the horizontal dimension, then, is the extent to which local officials should be relied upon to represent these diverse community interests. To a large degree, elected local officials have proven to be sensitive to public concerns about hazardous waste facility siting. In municipalities in which facilities have been proposed, officials frequently become spokesmen for the intense opposition to approval of such sites. According to the Centaur Associates study for EPA, local elected officials "almost always" oppose siting attempts, often by use of such tactics as resolutions against a facility, bans on acceptance of certain wastes, or orders to close a facility.[59] The basis for these actions accords with the best aspirations of the representative notion of democracy, whereby officials adhere to the will of the majority so as not to risk electoral defeat: "Because hazardous waste facilities are so unpopular, local elected officials who support them do so at the risk of their public careers."[60]

Given this apparently close correspondence between the dominant interests of a community and the actions of local officials, it is not surprising that little attention has been paid to the possible need to broaden community representation in siting policies to reflect a greater range of horizontal perspectives. This situation arises, though, because of the fact that a hazardous waste facility-siting proposal is generally viewed by community residents as a bad bargain in which the perceived costs far outweigh the perceived benefits. Instead of diverse local interests, in the case of hazardous waste facility-siting attempts one often finds a more homogeneous pattern of clear-cut opposition to the proposed new facility. Nevertheless, there is a chance that siting policies different than those pursued to date will be able to realign this "lopsided" predominance of local opposition to form a more diverse pattern like the one shown in Figure 4–2. Under these conditions, neither opposition nor support groups would necessarily command an overwhelming share of community allegiance.[g] To some extent, therefore, the ultimate test of a sound siting proposal and process is that it can bring forth groups in opposition, groups in support, and a body of local residents who are undecided.

When community interests are divided in their reaction to a siting proposal, of course, it becomes more difficult for local officials to represent all viewpoints. In energy facility-siting cases, for example, officials have tended to show a bias in their decisionmaking toward policies that are compatible with management of their community's budget. Local officials have often been more hospitable to facility-siting proposals than have the community residents, who weigh fi-

[g]A spokesperson for the Stablex Corporation, for example, contends that a small but vocal minority of local residents have been responsible for resistance to his company's siting proposal in Hooksett, New Hampshire. According to James McCoy, many Hooksett residents do not oppose the proposal. Thus it is difficult to speak of a single "community" given this situation.[61]

nancial benefits somewhat less heavily. In her analysis of energy facility-siting attempts in New Jersey, Grace Singer called attention to this differing viewpoint between citizen activists and local officials. While the former were averse to the prospect of greater health and safety risks posed by polluting facilities, the latter tended to be preoccupied with fiscal planning.[62] Singer noted that although the Dow Chemical Company's proposal to build a $30 million "tank farm" in Bordentown aroused negative citizen response, local government officials accepted the proposal due to the facility's financial benefits:

> Government officials were primarily influenced by the facility's estimated annual tax payments of about $1 million. They told citizens that the Dow facility would enable them to reduce homeowners' property taxes by $200 per family.[63]

Similarly, the siting of a municipal landfill in Jackson Township, New Jersey, resulted from town officials' acceptance of an attractive corporate land gift to the community (see Appendix A). The landfill was opened despite opposition from nearby residents and with little attention paid to its geological shortcomings. Subsequently, what the town officials had first seen as a "deal we could hardly beat" ended up contaminating groundwater supplies from toxic dumping of materials into the landfill, causing serious health impairments.

The mayor's office in Newark, New Jersey—a city facing severe financial difficulties—has lobbied in favor of locating a new hazardous waste facility in this community. At the same time, Newark officials argued that such facilities should pay a 10 percent gross receipts tax to the host municipality.[64] Those residents of Newark who live near a proposed site, however, may be less influenced by the facility's potential tax payments than by its potential threat to public health. This became apparent in 1980–81 as a proposal to locate in Port Newark a trans-shipment terminal for hazardous wastes headed for incineration at sea came under attack by local residents, particularly in the traditionally independent Ironbound section of Newark.[65]

This systematic financial bias on the part of local officials shows an important limitation of the representative model. The interests of local officials, by virtue of their positions as government employees and town planners, do not always correspond to those of the majority of community residents. Moreover, David Bradford and Harold Feiveson's concept of "legitimate discourse" is prevalent in such situations,[66] insofar as politicians stress quantifiable benefits while finding it more difficult to discuss unquantifiable or intangible costs such as fear and long-term public health risk.[67] Reliance on elected officials alone, therefore, does not provide an adequate guarantee that varied community interests will be included in the siting decision, particularly in cases where officials' perceptions of a facility's costs and benefits differ from the views of many of the citizens whom they represent.[68] Direct citizen participation in siting decisions is needed to ensure respect for the siting process.

The paucity of horizontal group participation in siting decisions in the basic tripartite model (developer, local officials, state officials), particularly at the community level, suggests the need for substantial refinement of this structure. The two additions made in New Jersey—the chemical industry, and environmental or public interest groups—while a step in the right direction, are not sufficient. Indeed, the fact that New Jersey's horizontal expansion took place only at the level of statewide interests indicates a subtle "stacking of the deck" toward those groups likely to favor state-controlled siting. Ironically, then, although the purpose of having extensive discussion of the siting bill's revision among the five statewide groups and of institutionalizing their input in future siting debates was to facilitate acceptance of whatever plan ultimately emerged, the actual distribution of interests in this decisionmaking process was weighted toward those who favored siting of new facilities, with less representation of those groups opposed to new sites whose acceptance of the siting process will be most critical to its eventual success.[h]

Such limited horizontal participation may well create an illusory initial consensus in decisionmaking. The parties who helped to revise New Jersey's hazardous waste siting bill, for example, appear to be in general agreement over the basics of this siting strategy, in large measure because those who will be most opposed to it have not yet entered the debate over state siting policy. Once actual sites are proposed in New Jersey, the opposition will undoubtedly be adamant, as has been the case in New York, Massachusetts, and elsewhere.

[h]Those who designed this process operated on the assumption that local interests could—and would—participate actively in the actual site review and facility application procedures set forth in the new statute. Strong public participation mechanisms were included in the process to accomplish this objective. Moreover, statewide environmental and public interest groups have made a commitment to oversee this process of local involvement, and to oppose any efforts by the developer or the state agency to manipulate local interests.[69] It remains to be seen how this theory will work out in practice.

The explanation for this delayed local opposition rests on the lack of structural strength among these groups who only mobilize when an issue becomes "salient." As Richard Greenwood explains, "Most people do not have the time or the money to become heavily involved, and usually become aroused only when the project affects them directly."[70] Clearly then, in order to deal successfully with such last-minute opposition, siting policies should attempt to incorporate a broader range of horizontal interests rather than limit local input to community officials. The ways in which this goal might be accomplished, and the difficulties involved in achieving it, require explanation of the structural dimension of siting interests.

THE STRUCTURAL DIMENSION

Opening the Door to Siting Interests

One answer to the question of who ought to participate in siting decisions is: "all interested persons." The principal drawback to this approach, obviously, is that the planning process would become far too ponderous and time-consuming. Nevertheless, as Murray and Seneker point out, "The risk does not seem as substantial . . . as the possibility of excluding persons with a valid interest."[71]

This tradeoff between administrative efficiency and what might be termed "participatory equity"—inclusion of *all* who have a stake in a decision—is analogous to the conflict between preemptive and override mechanisms described earlier. In much the same way that a preemptive siting mechanism strives for easy implementation of statewide objectives, the simple representative model offers an uncomplicated array of interests who might quickly concur on a decision. Both approaches unwisely settle on expedient solutions at the expense of greater long-term political acceptability.

It is not sufficient, though, simply to design a siting process that allows for an "open dialogue" and for input from "all interested parties." Such an aim seems rather naive. Many individuals who have a stake in siting, for example, are unlikely to participate in even the most open process due to their lack of awareness of the paucity of information available to them. According to R. Kenneth Godwin and W. Bruce Shepard:

> . . . it is quite probable that the vast majority of the persons who will be adversely affected by state land-use policies, as they are presently emerging, are presently unaware of the importance of the issue to them.[72]

This problem suggests the need for adequate publicity about siting plans in order to make the public aware of the plans' potentially important impact. Hearings on the New Jersey siting legislation by the Senate Energy and Environment Committee repeatedly exemplified this informational problem. The few citizens who attended these hearings who were not already members of organized groups complained at times about the lack of publicity given to the original siting legislation and to the Senate's process for revising the bill.[73]

Even if individuals were aware of the siting process and understood its potential to them, structural differences prevalent among siting interests would nevertheless give rise to participatory discrepancies. Many of the groups shown in Figure 4–2, while potentially having a large interest in siting, are not organized to advance such interests. As the process of revising siting legislation in New Jersey illustrated, those with power over government decisionmaking are already *organized*. In fact, the environmental and public interest groups enhanced their influence in this process by forming a twelve-group coalition in order jointly to recommend changes in the bill.[74] Hence, an important dictum in siting appears to be that "Not all persons will be represented . . . because they do not belong to an organized group that is politically active on the specific issue."[75]

Recognizing the unlikeliness of "getting people off the street" to participate in decisionmaking on hazardous waste facility siting, some believe that the involvement of various organized groups will provide sufficient "public" input. For example, Diane Graves of the New Jersey Sierra Club asserted that if any 500 individuals were to participate in a siting process, they would select leaders and sort themselves out into groups. Rather than taking the time and effort to have this "sorting out" process occur, however, she considers it easier to rely on existing organizations at this stage. In turning to these groups, Graves feels that one can "safely assume" that the different siting interests will be adequately represented. Local participation will occur later, in the debate over actual siting proposals.[76]

Who's Missing From This Picture?

A process that restricts participation to organized interests, however, necessarily creates a substantial bias against those interests which have the least structural power. As Figure 4–2 shows, at least five important groupings of this type exist: state or regional support, industrial hazardous waste generators, regional opposition, local support, and local opposition.

The members of the public-at-large are least likely to participate directly in any siting process. Many of them, however, can generally be included within the "state support" category, since they would benefit from safe management of wastes and from lessening of illegal or unsafe disposal. For this group, siting of new facilities represents a large collective benefit but a small per capita benefit. The interests in siting of this large body of unorganized citizens traditionally have been ignored by systems which rely on local decisionmaking, to which these individuals have almost no access.[77] At present, the interests of the ordinary state citizen are indirectly represented (if at all) by state agency officials, or by representatives of environmental and public interest groups, the business community, the chemical industry, or organized labor.[78] The movement toward preemptive or override state siting mechanisms is intended, among other goals, to incorporate the interests of this larger group of citizens in adequate waste disposal.

Generators of hazardous waste are another statewide grouping of unorganized interests. This group, too, is a diffuse constituency. In New Jersey alone, for example, there are as many as 10,000 generators of hazardous waste.[79] Nationwide, over 50,000 firms responded to EPA's initial RCRA filing requirements for waste generators.[80] The largest generators—giant petrochemical companies—are easily identified. Accordingly, New Jersey's chemical industry, through those companies which are members of the New Jersey Chemical Industry Council, participated in redrafting the state's siting bill. In having the major chemical companies in the state speak for the category of all waste generators, however, the legislative process again adopted a strategy somewhat contrary to its own professed goals. The most ludicrous aspect of relying on large chemical firms to represent the interests of waste generators as a whole is that these particular generators are the ones *least* in need of commercial off-site facilities; the vast majority treat their own wastes on-site. Construction of new off-site facilities, however, is the obvious aim of the New Jersey legislation.[81] Nationwide, the large petrochemical companies have made relatively little use of commercial facilities, disposing of 93 percent of their wastes on-site.[82] This illustrates that the firms which are organized and can most easily influence siting policies are not necessarily the greatest potential users of any new facilities sited. The large chemical companies should therefore not be the only ones to speak for waste generators as a whole. Their participation in the process is desirable; their domination of it is not.

A desirable siting policy should undoubtedly attempt to include a wider segment of the waste-producing sector. Without such participation, facilities sited may not necessarily be put to use, which is one of the worst outcomes possible. The lack of a clear generator-disposer relationship has helped thwart siting attempts, since the developer of a facility often has had to "take the heat" alone, rather than being backed up by future clients of his proposed facility. From the standpoint of an effective siting program, it is advantageous to make this relationship as evident as possible. The siting of on-site facilities, by definition, shows where the waste will be coming from; perhaps partly for this reason, on-site proposals have tended to be more successful.[83]

Inclusion of a wider segment of the waste-producing industry in the siting process will not be easy since many of these generators are small businesses. One possible strategy is to seek participation by a representative sample of all generators. The information thus obtained will be a necessary component of sound siting policy, particularly in confirming the demand for new facilities. The manifests required by RCRA might be used to identify those generators who will need additional storage and disposal capacity.

Regional interests in siting are a third grouping whose input in siting decisions has considerable importance. While the general public of average citizens has a low individual stake in siting, residents of the region can suffer potentially severe impacts. It is widely acknowledged that the costs of new hazardous waste facilities do not respect political boundaries. Yet most siting policies established to date have not dealt with this problem in a satisfactory manner.

The importance of including representation in siting policy from nearby communities has been recognized in the Connecticut and Massachusetts legislation, among others. Connecticut's statute provides that the local project review committee which negotiates with developer-applicants must include at least one elected official from a nearby municipality.[84] Likewise, the Massachusetts Hazardous Waste Facility Siting Act allows the chief executive officers of abutting communities to negotiate with developers over compensation. Agreement is a necessary precondition for approval of a developer's proposal. Disputes are to be settled, if necessary, by an arbitration panel.[85] Nearby local governments can probably represent regional interests reasonably well since the citizens affected comprise a structurally weak collectivity. Hence the reliance on community officials in the Connecticut and Massachusetts legislation seems a good approach, particularly since these officials are likely to be acutely conscious of any uncompensated costs being imposed upon their communities by the nearby facility.

Inclusion of local support and, especially, of local opposition (which, in the case of hazardous waste facility siting, is likely to greatly outnumber support) is another necessary feature of any effective siting policy. Siting impacts bear most heavily upon local residents. Moreover, as suggested earlier, the interests of local officials are not always coextensive with those of their constituents. The abutters group shown in Figure 4–2, for example, will be most directly exposed to whatever pernicious impacts are associated with operation of the new waste facility. Thus, these residents have the most reason to protest the building of this facility in the absence of sufficient offsetting benefits.

Because of the importance of siting to local residents, many believe that those most heavily impacted should be specifically included in the decisionmaking process rather than being represented solely by their local governing body.[i] The National Governors' Association, for example, recommends that participation by local special interest groups, particularly abutters, be encouraged.[87]

Because a waste facility's impacts within a community will be so high in per capita terms, and so highly visible, a common assumption is that structural groups from the locality will emerge to participate

[i]One difficulty is deciding *when* to supersede the representative model. One possibility is that the larger the magnitude of the facility's impacts on the community, the more individual community interests should partake of the decision. This notion seems to be borne out by the use of referenda in other siting attempts which appear to favor participatory democracy in instances when it most "counts." There was a trend in the middle 1970s toward use of referenda in New England states to determine local attitudes toward energy facility proposals.[86]

in siting decisions. Adequate public participation, according to Diane Graves, for example, then becomes a matter of including those groups within the siting process.[88] However, a definite danger is associated with this assumption. People who participate actively in political processes tend to have sufficient resources to allow them this "luxury." Hence, political participation tends to be characterized by an unmistakable bias against lower socioeconomic groups[89] and against those lacking effective organization. As Godwin and Shepard cautioned,

> The persons who are unorganized are not likely to be represented. Since the unorganized are often those persons not endowed with the economic and political resources necessary to become organized, regulatory politics generally exclude the poor.[90]

This bias inherent in the structural dimension must be carefully considered in any siting policy, especially since it goes hand-in-hand with society's tendency to impose burdens on politically and economically disadvantaged groups.

Involvement of local interests meets two objectives: pragmatism and equity. In many cases, exclusion of these groups will intensify resistance to the proposed facility. Moreover, it seems fair that those who will be most affected by a facility have a voice in decisions about its location and construction. When groups do not automatically emerge to protest facilities, participation by area residents should not be excluded. To do so may encourage a not-so-subtle tendency to put these facilities in places where people ("the poor") will not object to them, rather than in places where they will do the least environmental damage.

However, citizen groups in at least some poorer, environmentally-degraded urban areas are beginning to organize strong resistance to proposals to site potentially dangerous facilities in their neighborhood. Citizens in Jersey City, Bayonne, and Hoboken, New Jersey, successfully defeated five siting proposals in the late 1970s;[91] resistance to a transfer point for at-sea incineration of hazardous wastes has emerged from the Ironbound section of Newark, New Jersey;[92] and the impoverished residents of Starr County, Texas, successfully opposed a hazardous waste facility.[93] Nevertheless, concerns remain about the ability of groups from poorer communities to withstand powerful siting pressures.

The preceding discussion, in referencing three main dimensions of the siting public, has pointed to some of the major considerations involved in designing an adequate siting policy. Having identified "the public," the next step is to focus on different rationales for public participation.

PUBLIC PARTICIPATION: PRELUDE TO DECISIONMAKING

Proposals to site hazardous waste facilities almost inevitably stress the importance of public participation procedures. The apparent consensus on this point is alarmingly deceptive, however. For while references to public participation cater to the American ideal of democratic decisionmaking, behind this common approbation of public involvement lie very different conceptions of the purpose of participation and of its value within the siting process.

Public Approval: The Elusive Search for Consent

Obviously, one of the major purposes of involving the various interests in decisions on new sites and new facilities is to enhance people's acceptance of the final decision. The Chemical Manufacturers Association's position on siting legislation and public participation, not surprisingly, focuses on this aspect:

> The most important ingredient of successfully attaining sites for hazardous waste disposal facilities is that of involvement. The local people, local government, local thought leaders, and local press must be thoroughly informed of all facts as early as practical. These local people should be part of the decisionmaking process.[94]

In CMA's "model" siting legislation, this involvement is to be accomplished largely by means of public notices and hearings.[95] The association's incorporation of certain basic participatory measures into its proposed siting process represents an attempt to enhance the legitimacy of the siting process by allowing local citizens to have a voice, albeit limited, in these decisions. Public support is seen as the "key ingredient" in successful siting.[96] A similar perspective toward public participation was offered by the National Governors' Association's (NGA) final report on siting hazardous waste facilities:

> The rationale for participation is to achieve timely decisions on facility proposals that will result in construction and successful operation of acceptable facilities.[97]

The word "acceptable" in this NGA policy statement is quite important. Public participation mechanisms are intended by the NGA to improve the quality of siting decisions, not only by rejecting ill-conceived proposals but also by modifying any proposal to better take account of the various interests it affects.[98]

Involvement includes participation in creating the basic siting legislation, participation in developing siting criteria and needs assessment, and participation in applying those statewide criteria to selection of actual sites for facilities. As the New Jersey Sierra Club's Diane Graves summarized in a November 1980 discussion:

> Without adequate, effective public participation at all stages of the process, siting will simply be impossible. People will refuse to agree to facilities proposed for their communities. This is the prerequisite to everything else.[99]

A basic strategic lesson gleaned from past siting failures is that the public must usually be informed early in the siting process. The EPA report on public opposition notes:

> One factor in particular which has been blamed for the demise of a number of sites and potential sites is the failure to inform local residents and elected officials of development plans, so that they are presented with a *fait accompli* in terms of site location and facility plans.[100]

One of the most agreed-upon propositions of siting has thus become to "involve the public early."[101]

Involvement is not so simple, unfortunately. Sherry Arnstein has identified eight distinct levels of citizen participation:

- citizen control;
- delegated power;
- partnership;
- placation;
- consulation;
- informing;
- therapy;
- manipulation.

The bottom two categories correspond to "nonparticipation"; the next three, to "tokenism"; and the top three, to increasing amounts of "citizen power."[102] The strategic conception of public participation, prevalent not only in industry but among certain government agencies as well, clearly focuses on the bottom half of Arnstein's typology.[103] The very bottom category, "manipulation," embodies the essence of these strategic concepts since its emphasis lies wholly on utilizing public participation as a public relations device. The manipulative approach frequently backfires in controversial cases, however, leading to "deep-seated exasperation and hostility" by citizens toward those making the decisions, and to frustration on the part of facility developers.[104] This is exactly the opposite of popular respect for the siting process.

Successful strategy dictates resort to measures above nonparticipation. Much of the rhetoric of public participation, and many of the formal state and federal participatory requirements embedded in such statutes as the Clean Water Act, Housing and Community Development Act, and Clean Air Act have been failures to date. These approaches have glorified formalism over true public influence on agency decisions. Furthermore, the information presented has often been slanted toward the desired results of the hearing.

Many observers are optimistic about the chances of doing better with respect to hazardous waste management, an optimism based at least in part on their perception of its necessity. "We have no choice but to involve the public," they typically argue. "Without public involvement we won't get any sites approved." The challenge to government and industry to accommodate public participation is enormous, however, as is the challenge to individuals and groups to participate both responsibly and positively, rather than just emotionally and negatively. Paternalism by either industry or government agencies—slick advertising, disdain for ordinary citizens—will simply lead to the rejection of one site after the next.[105]

The public notice and hearings provisions embodied in CMA's model siting statute and in many state laws on siting ("tokenism," according to Arnstein) will not be sufficient. Reliance on these mechanisms falls within the "consultation" level of the participatory hierarchy:

> Inviting citizens' opinions, like informing them, can be a legitimate step toward their full participation. But if consulting them is not combined

> with other modes of participation, this rung of the ladder is still a sham since it offers no assurance that citizen concerns and ideas will be taken into account. The most frequent methods used for consulting people are attitude surveys, neighborhood meetings, and public meetings.
>
> When powerholders restrict the input of citizens' ideas solely to this level, participation remains just a window-dressing ritual. People are primarily perceived as statistical abstractions, and participation is measured by how many come to meetings, take brochures home or answer a questionnaire. What citizens achieve in all this activity is that they have "participated in participation." And what powerholders achieve is the evidence that they have gone through the required motions of involving "those people."[106]

All strategic approaches to participation fundamentally involve the quest for legitimacy. Their intent is that citizens (by having participated in participation) will view the siting process as legitimate and thus will accept its outcome. Legitimacy, according to Burke, results when individuals sanction the intervention and influence of an organization; to obtain legitimacy is thus akin to obtaining permission to take certain action.[107]

The greatest shortcoming of the strategic approach arises when siting processes with only token participatory measures are not seen as legitimate. This type of response may, in fact, prove common, as exemplified by the public reception to the original version of New Jersey's siting bill. Despite the assertion of the bill's legislative findings section that "public participation should be built into every step of this planning and siting process,"[108] the public participation measures were to be confined to exactly the types advocated by CMA: public notice, hearings, and local minority representation on the siting board.[109] Local officials and environmental and public interest groups strongly opposed this proposed legislation. The executive director of the New Jersey State League of Municipalities, for example, termed the original bill "grossly unacceptable," largely because of the absence of mechanisms allowing for adequate local and public input.[110]

Thus, the consultation approach—which many, like CMA, perceive as a means of obtaining public support for siting measures—is instead commonly viewed by citizens as a "sham." Such measures provide no assurance that citizen input actually will have any effect. In contrast, the public needs to know that its concerns will be (1) heard and (2) really taken into account. Although public concerns might not always be agreed to by government and industry, they should always be a major component of the final decision. If these two features of participation are met, many people will respect the siting process and perhaps even accept its eventual results. Much more is involved than simply providing for public hearings after industrial and governmental siting decisions have already been made via the traditional "decide, announce, defend" model of siting.[111] Instead, the process must create opportunities for an intense local involvement as well as access to state override decisions.

Public participation is essential in the development of new siting legislation. In Michigan, for example, participation in devising toxic substances control legislation was apparently quite successful—after the PBB disaster provided a sharp catalyst.[112] The changing currents of government interaction with the public in New Jersey in 1980–81 provide another example. Here the participatory process did improve substantially, though the effort took much time and witnessed many disappointments as well as successes. In 1979 Governor Brendan Byrne appointed a Hazardous Waste Advisory Commission composed of representatives from industry, local and state government, universities, and one environmental organization. At the same time, the New Jersey Department of Environmental Protection (DEP), jointly with the Delaware River Basin Commission (DRBC), was developing tentative siting criteria for the state. In January 1980 the Governor's Commission issued its report, which recommended, among other things, that the public be involved in developing new state siting legislation. Instead, the state's Department of Environmental Protection went ahead and, in secret, devised a statute on its own, which was soon thereater introduced in the State Senate as S. 1300. Critics reacted sharply, arguing that the bill was grossly inadequate and that the public had been excluded from the legislative development process.[113] Not long thereafter, a stormy public meeting took place in East Windsor Township, where citizens were upset that the DEP-DRBC siting criteria had made their town a likely candidate for a new hazardous waste facility.

All of this controversy led key legislators, particularly the chairman of the Senate Energy and Environmental Committee, Senator Frank Dodd, along with members of the state administration, to commit themselves to a different approach, one in which concerned members of the public—environmentalists, local government officials, and industry—would be able to participate. The Senate Committee convened a series of effective meetings on S. 1300, and the bill was entirely rewritten prior to its final legislative approval in mid-1981.[114]

Public participation is also important in developing siting criteria. It is essential that concerned groups and individuals in the state come to feel that they have had a fair opportunity to influence the selection of criteria: technical criteria (soil characteristics, floodplains, groundwater aquifers, and so on), and sociopolitical factors (land-use compatibility, population proximity, transportation routes). If participation is ineffective at this stage, it may become impossible to gain trust later, when actual sites for new facilities are proposed. States need to reach out to a wide range of people at this juncture and not wait until later when specific sites are proposed. Conversely, if many people come to perceive the state's final siting criteria as sound results of an open debate, then opposition to later use of these criteria to select actual waste disposal sites will be lessened (though not eliminated). Opponents at this later stage will be somewhat isolated. Early public participation can therefore help build a powerful constituency for the overall process. If citizens throughout the state have become convinced—through effective participation—that new treatment, storage, and disposal facilities are needed and that the explicit siting criteria are indeed the best ones to employ, then selection of a site even in their own community will have a strong logic.[115] Opposition will still emerge, however—siting criteria, like public participation mechanisms and balanced decisionmaking authority, are no panacea for successful siting.

In general, the strategic approach to participation has tended to embrace what Lawrence Susskind has labelled "advice-giving strategies":

> Advice-giving strategies are generally aimed at building support for proposed policies by offering stakeholding groups a chance to express their values before policies are adopted. The key assumption underlying this strategy is that advisory groups that have been given a chance to help shape a policy will be more likely to support the policy after it is adopted.[116]

As the opposition to the original version of S.1300 in New Jersey demonstrated, reliance on advice-giving and other "token" participatory measures is insufficient to achieve the desired acceptance of siting policies, particularly to the extent that these mechanisms are seen as embodying what Arnstein has termed "window-dressing ritual."

Participation and Improved Planning

Public participation measures need not be designed merely to gain approval and support for a particular policy. Another purpose widely recognized in the literature on public participation consists of using participation to provide a source of additional data and opinions, thereby improving the planning process.[117] Discussion by a wide range of interests can improve decisionmaking by calling attention to possible errors in proposed plans and suggesting significant revisions. Frequently, these improvements consist of establishing ways to reduce the risk of a project to a more "acceptable" level. Alvin Weinberg has described how public participation improved the decisions made on nuclear reactor safety:

> . . . in the Soviet Union, where the public does not have an automatic right to be informed about or to participate in scientific and technological debate of this sort, the technology of reactors is rather less obviously centered around safety. Until recently, Soviet pressured water reactors had no containment shells. . . . There was here a divergence between the American and Soviet views, both with respect to the effectiveness of containment shells, and with respect to how safe is safe enough . . . In my view, the added emphasis on safety in the American systems is an advantage, not a disadvantage; and insofar as this can be attributed to public participation in the debate over reactor safety, I would say such participation has been advantageous.[110]

Use of public input to suggest improvements in decisionmaking reflects an approach that Susskind labels "error detection":

> Error detection strategies presume that even with the advice of stakeholding interests, policy designers (i.e., elected and appointed officials) may misinterpret facts or overlook considerations of special importance to certain groups.[119]

Public participation could contribute to error detection in two ways. One, as Susskind suggests, is simply as a check on the technical accuracy of the planning and an opportunity to consider additional relevant factors. (Risk analyses, for example, may exclude some important accident scenarios.) As the Keystone conferees noted: "Public participation is a tool to see that the decision is rational and based on best available information."[120] Public participation, however, can also help infuse the planning process with representative values and political goals. Insofar as decisions on sites for hazardous waste facilities are "trans-scientific," the underlying scheme of values on which they are made has no "objective" validity.[121] An error-detection strategy, therefore, should contribute political as well as technical insights to decisionmaking, allowing open questioning of the possibly undesirable assumptions that lurk behind siting decisions.

Information dissemination and response to citizen queries are key elements of this process. The information needed is not advertising or propaganda, but honest explanations in language that people can comprehend. Since people so often have an image of Love Canal or of another landfill in mind when they hear the phrase "hazardous waste disposal facility" or "hazardous waste site," the first priority in information dissemination involves describing the facility's actual characteristics. The popular response to a proposed new petrochemical facility is different from the current response to a hazardous waste treatment facility or high-temperature incinerator, even though the plants actually may be rather similar. Information can help clarify this distinction in people's minds. At the same time, if a landfill of some kind is contemplated, this fact must be presented at the beginning, honestly and openly.

The chemical industry, waste management firms, and government regulatory agencies would help reduce opposition to siting if they acknowledged past problems of managing hazardous waste and communicated their commitment to do better in the future, providing specific information on new technical and managerial approaches. More than mere words will be needed, however—they will have to demonstrate on a sustained basis that they can operate or regulate such dangerous facilities in a responsible manner. People's perceptions of their actions in this regard are probably more important than reams of technical information or slick slide-and-tape audiovisual presentations. In this sense, for example, actual photographs may have greater credibility than would artists' drawings.

Information may also be needed on the origins of the wastes to be treated or stored at the site: Will they come from the local area, from far away, or from both? Another important issue involves the types of wastes to be handled. People want to know if "political wastes"—PCB's, kepone, dioxin, and so forth—are to be present. Such wastes automatically raise the visibility of the debate due to widespread public recognition of their harm. Moreover, people want to feel secure that no significant changes in waste composition will occur after the facility begins to operate, unless they have a say in the matter.

Understanding the need for the facility is another obvious theme in the participatory process. This relates the individual siting decision to the broader state inventory of waste treatment needs and existing capabilities. "Need" may have to be subdivided into generation of new wastes and cleanup of existing dumps. People will be somewhat more prepared to accept a facility proposed for their community if they are convinced that it is needed and that the proposed site is really adequate.

Protecting Rights Through Participation

Beyond gaining public approval for decisions or improving planning, public participation can also help protect individual and collective rights.[122] As the report of the first Keystone conferees affirmed, "Used properly, [public participation] is a tool to help assure that decisions are wise, just, and fair."[123] Participatory measures thus offer a means to reconcile policy decisions to the tensions between equity and efficiency and to the conflict between majority rule and minority rights.

Those adversely affected by a hazardous waste facility are particularly prone to complain that the decisionmaking process was fundamentally unfair. Local residents (the "local opposition" shown in Figure 4–2), if given a chance to voice their objections through a participatory format, will be able to speak out against actions of others that threaten their own interests. Without local input, decisionmakers often will not be aware of the full range of costs—including political costs—associated with a particular project. Indeed, unquantified costs traditionally have been susceptible to insufficient weighting in public decisionmaking. Residents of a community may be able to point to the impacts associated with a proposed site that have not been adequately accounted for, as in the case of the Calabasas landfill in Los Angeles which posed substantial aesthetic costs to nearby residents. By being able to make such concerns known through public participation procedures, individuals are not only fulfilling the purpose of "error detection" but are also asserting their right to have their interests incorporated in the final decision.

Susskind's third strategy of participation, negotiation, corresponds to this aim of protecting individual and group rights. Negotiated approaches, he has said, ". . . assume that all stakeholders (including policy-makers) must share responsibility for reconciling competing interests."[124] The strongest complaints about the injustice of a given policy would likely result from those individuals and groups which were not even allowed to enter into the decisionmaking process.

Limits to Participation

Many hazardous waste management facility siting attempts avoid true participation, hoping thereby to achieve success in site approval. The EPA-sponsored study of local opposition, for example, concluded at one point that:

> Under certain circumstances, such as when the site is in a heavy industrial area and not in the public view, a low-profile approach may be warranted. There is at least some evidence that opposition will not arise in these cases, so that there is no need to alert the public and thereby create a potential for opposition.[125]

The report indicated that this "low-profile approach" had worked in cases such as the siting of a 200-acre landfill in Texas City, Texas. The Gulf Coast Waste Disposal Authority had argued for the desirability of this approach:

> . . . the public could not be expected to willingly accept such a site simply because it understood the problem. Furthermore, they felt that such a program might be counterproductive, since it might create opposition where none would exist otherwise.[126]

The EPA report also noted that publicity had been "studiously avoided" by Frontier Waste Processing in this company's siting of a facility in Niagara Falls, New York.[127]

The view that in some instances "there is no need to alert the public" is an especially dangerous one. Even by a strategic logic of siting, this approach carries a risk of backfiring and creating intense opposition by virtue of the fact that participation has been circumvented. Indeed, this narrow, naive philosophy, if adopted by the states, could eventually intensify public opposition to siting hazardous waste facilities anywhere.

More importantly, however, the "low-profile" approach runs directly counter to the goals of protecting rights and improving planning through effective public input. "Silent siting" poses a special threat to the rights of residents near a proposed site. By choosing not to inform them of the future existence of a hazardous waste facility which may adversely impact them, decisionmakers deny these residents an opportunity to protest the imposition of these ill effects and perhaps to improve the design of the facility to lessen its impacts. While putting hazardous waste facilities in some citizens' "backyards" may be required for the larger benefit of society, it is fundamentally wrong to impose these costs without adequately informing those who will be affected.

The ethical issue of knowledge versus consent is by no means trivial. While local consent may not be required legally for the state to take an action, it is eminently reasonable to require the state to provide prior knowledge of these actions. Such information is necessary, from an ethical viewpoint, inasmuch as what one doesn't know *can* hurt when one is dealing with hazardous waste, given the history of abandoned dumps.

HOW TO DECIDE? NEGOTIATION AND SHARED AUTHORITY

Judging from their results, local home-rule decisionmaking structures in effect in many states have not provided an adequate answer to the question of how best to decide on sites for hazardous waste facilities. Much time and effort typically is invested in trying to locate a facility, yet the end result almost inevitably is either the facility's outright rejection or unavoidable delays that force the developer to withdraw his proposal.[128] Even worse, though, is the fact that new policies being devised by states across the country to deal with this problem—primarily through preemption—may not work well either.

Tom Cook and James Knudson maintain that "Some method of selecting and involving the local community must be found that does not lead to the familiar cycle of protest and opposition."[129] When set against a background of repeated failures to obtain sites for needed facilities, it is easy to see why priority is being given to the *successful* siting of facilities, with less attention paid to considerations of comparative efficiency and equity. In contrast, the basic challenge is to arrive at a decisionmaking mechanism that not only works but that also satisfactorily incorporates the legitimate interests of all those affected. From the perspective of decisionmaking authority, creation of a balanced, sequential, and timely process has been identified as especially desirable. What does this suggest for participatory mechanisms?

The advice-giving strategy appears far too limited to warrant its use as a basis for siting decisions. In attempting to ratify decisions through use of limited public input, it settles for "token" participation which not only may prove to hamper its own effectiveness but unfairly deprives many siting interests of an effective voice in those decisions crucial to their safety and welfare. Perhaps the biggest danger of a strategic advice-giving approach is the possibility of instilling increased feelings of political alienation among individuals who realize that their role in siting makes no real difference. As Donald Mazziotti explained, "Feelings of alienation grow out of the perception that political decisions are made by a group of political insiders who are not responsive to the average citizen."[130]

The error-detection method corresponds to a more beneficial purpose of participation: improved planning. The basic mechanism, however, revolves around use of public hearings, a mode of public participation with obvious shortcomings.[131] The link between environmental policymaking and the public hearing mechanism has been well established within the past decade. Use of public hearings tends to be associated with a decisionmaking process in which a facility developer (1) submits an application to a government agency, which (2) conducts a review of the application prior to permitting (for example, preparing an Environmental Impact Statement), and then (3) before licensing the facility, holds a public hearing to expose the plan to public scrutiny. Following this process, a permit is either approved or denied by the agency.[132]

While public hearings frequently have been used in environmental decisionmaking, this has not always proven to be desirable. As a result of its mandated use in many governmental programs,[133] the public hearing has tended to become merely a "ritualistic" device whose purpose is often not so much to expose errors in the proposal as to validate the formal decisionmaking process. Hearings tend to be devoid of fundamental political dynamics and are all too often the last stage of the paternalistic "decide, announce, defend" (DAD) sequence whereby the agency involved finds itself in the position of justifying to the public a decision it has already made.

Though public hearings have been used largely as a formal device to present a proposal to the public, this mode of participation could be potentially valuable. Hearings could be divided into two stages: the first a series of information sessions at which the proposal is outlined to the public and issues of local interest are determined; the second set of hearings, in contrast, would allow for precise debate of the relevant policy issues and would thus employ a more adversarial framework.[134]

Participation in informal workshops, seminars, and evening discussion sessions can often be valuable in achieving popular understanding and eventual consensus. Informal discussions with state environmental activists, among others, could expand public knowledge and awareness.[135] Thereafter, the public could meet more productively with government and corporate officials. Use of meetings and hearings in this manner would also tailor the decisionmaking process to the "trans-scientific" nature of siting, by first identifying the technical, objective issues and then arguing about the underlying political problems.[136]

Clearly, even two-stage hearings are not enough. The EPA-sponsored report on opposition to siting described the limitations of hearings in decisions on hazardous waste sites, noting that hearings usually have ". . . provided for only the most limited substantive discussion and primarily served as forums for expounding positions for or against proposed or operating facilities."[137] According to this report, a public hearing held during the successful siting of Monsanto's landfill in Bridgeport, New Jersey, had little impact on public acceptance; rather, the hearing was but "an administrative ritual" since the township had earlier decided to support Monsanto's plan.[138]

Ballot box initiatives and public referenda are controversial but potentially useful techniques for gauging public attitudes on a facility proposal. Obviously they pose great difficulties: the protagonists may oversimplify issues, manipulate voters through misleading claims, or spend a great deal of money on advertising.[139] If the ballot question is not worded very carefully, confusion may grow.[140] The timing of a referendum, too, may influence its results. Moreover, by their very nature referenda make it impossible for individuals to display degrees of preference, and may be an impractical way to ascertain true local wishes. By polarizing the debate into a yes or no proposition, it may become more difficult to organize the public debate around the question of "how could we modify this proposal to make it acceptable?"[141] Nevertheless, when state override of local opposition to a hazardous waste facility is contemplated, tensions are high and much is at stake. A nonbinding referendum to ascertain public attitudes on the siting proposal might be warranted. Local elected officials and members of the state legislature would then know better how their local residents feel about the proposed waste facility.[142] At the very least, use of a referendum should be considered by local decisionmakers.

State regulations for control of the waste management process, and for monitoring these facilities' impacts on the environment, also require public participation. Local residents must come to understand how these regulations are supposed to work and how they have been applied elsewhere. In this segment of the debate it is particularly vital that state officials be realistic, acknowledging the risks of inevitable leaks from landfills, the possibility of accidental spills, and so on, while explaining what can be done to control the overall risks. Both the level of risk and the proposed response strategies need to be communicated.

One innovative approach to extending participation involves training (by the facility developer or the state) of several local residents in basic techniques of environmental monitoring: periodically testing ground and surface waters in the area for possible contamination. Citizens will feel more secure if they can on their own supplement industry or state government assurances that "all is well" with their own monitoring.[143]

The limitations of public hearings reflect a problem with these decisionmaking processes that extends far beyond the fact that they are often too formalized and do not live up to their potential to improve planning. An error-detection approach, no matter how sensitive, is basically a process which seeks to invest the power of decisionmaking in a concentrated source, typically a state or local government agency. The French political system illustrates this type of concentrated power in a fairly extreme form. In developing nuclear power, the French decisionmaking system has placed very little reliance on meaningful public input. Furthermore, as Dorothy Nelkin and Michael Pollak explain, it has arrived at decisions which some citizens strongly oppose:

> In general, France's highly centralized political system provides few channels for effective public input on national policy. And on nuclear power in particular, aside from the pro forma public hearings required to obtain a construction permit, local or regional groups have no mechanisms by which to insert their objections into the decision-making process.[144]

The French system, with few effective "checks and balances,"[145] stands in contrast to the American political tendency to limit endowment of power to any single source for fear that such power might too easily be abused.[146] The fact that the existing political system has been largely unable to site new hazardous waste management facilities provides positive affirmation of the existence of genuine constraints on the use of power in the United States. Indeed, some might argue that the American political process lies too far on the opposite of the French one, insofar as it widely distributes veto power at the price of allowing desirable policies to be put into practice.

It is not surprising, then, that siting "reforms" in the United States frequently have focused on preemption. By establishing a "super siting agency" political power can be redistributed into the hands of a smaller body entrusted with the task of finding sites. Obviously, the power of those opposed to specific site determinations would be reduced by this step. Nevertheless, such hopes for an effective consolidation of political power over siting seem unrealistic. The range of opposition tactics available to opponents of facility siting is so wide that "streamlining" is an elusive, futile quest. In this regard, constraints on unpopular political action are so firmly entrenched in America that even an explicit legislative attempt to circumvent opposition is unlikely to prove successful.

The possibility of lengthy judicial review is, of course, one of the best weapons in the arsenal of the opponents of facility siting. Decisions made under either the advice-giving or error-detection models of participation will almost certainly be appealed to the courts. Despite the desire to endow a state siting agency with the final say in site determinations, the ultimate decision thus would frequently be in the hands of the court.

This model of decisionmaking, of course, is common within the American political sysem. By checks and balances, potential losers can guard against enduring these losses. Indeed, they frequently resort to the long, drawn-out judicial process to protect their interests. The result is a system that Lester Thurow labels "paralyzed," in which decisionmaking continually ends in a stalemate.[147]

Political power, then, appears to be distributed so that decisions about where to put hazardous waste facilities are potentially unenforceable. The critical question, though, is not *whether* it is possible to alter this balance of domestic power, but instead whether it is even *desirable* to do so.

The Negotiated Solution

One approach to siting hazardous waste facilities that is gaining greater currency involves dealing with opposition, rather than trying to bypass, overrule, or preempt it. A report by Clark-McGlennon Associates to the New England Regional Commission, for example, suggested:

> Opposition to these [hazardous waste management] facilities should *not* surprise us. Rather, this local response to a regionally beneficial facility should be anticipated and addressed, not ignored or fought. The opposition will not go away if ignored or slighted.[148]

The objective of dealing with opposition can best be achieved through Susskind's third category of participation: negotiation. Its implementation should be in the context of authority mechanisms that are balanced, sequential, and timely.

Public opposition to the siting of facilities arises out of well-grounded fears of the impacts associated with their operation. Through negotiation, these concerns are dealt with as part of the siting process, and their resolution is an essential component of successful siting. A report by Booz, Allen & Hamilton on New Jersey's prospects for developing sites found that:

> In the past, there have typically been only limited attempts to identify the specific nature of the objections of a community to a proposed facility and hence little opportunity to develop specific solutions to these problems.[149]

Instead, the authors suggested that negotiated siting policies could identify the specific objections of local opponents to a proposed facility and develop "reasonable, acceptable solutions" to these problems.[150]

Initial efforts have been made in New England to establish negotiation as a workable means of siting hazardous waste facilities. In contrast to the traditional adjudicatory approaches, here different siting interests are involved as "voluntary participants in a consensus building process." Negotiation, according to a study sponsored by the New England Regional Commission (NERCOM):

> is distinguished from an adjudicatory procedure for settling disputes since the parties work together in the problem of definition, in the design of the alternatives, and in the choice of a mutually acceptable decision.[151]

In both Massachusetts and Connecticut, legislation has been adopted which provides for a negotiated siting process. In Massachusetts, a necessary prerequisite to approval of a facility is the completion of a siting agreement between the developer and the host community's local assessment committee.[152]

It is possible, however, that negotiation or mediation (in which a third party helps to resolve disagreements) may prove unworkable in the case of hazardous waste facilities. Communities that vehemently oppose the siting of such facilities are unlikely to be willing to sit down with a developer and work out acceptable conditions for hosting it. This apparent drawback to negotiated decisionmaking weighed heavily in the analysis by Booz, Allen & Hamilton:

> First, while mediation can help in focusing attention on specific components of a conflict, it does not assure that there will be acceptable solutions to those problems. Secondly, there are no guarantees that opponents of a proposed facility will agree to mediate the issue or that these opponents will see any room for compromise regarding the facility. For example, if a local community feels that there are insufficient guarantees that a facility will not prove a threat to human health or the environment, no amount of negotiation will overcome citizen opposition.[153]

The obvious disadvantage of negotiation in hazardous waste facility siting is thus its lack of coercive power. Given the enormous risks involved, proposals to build facilities are not likely to be embraced by most communities; better instead simply to refuse to negotiate, or at least to refuse to negotiate in good faith. Inasmuch as the siting problem involves a conflict between majority rule and minority rights, negotiation appears to give particular advantage to the minority, at the possible expense of the majority's interest in successful siting.

One answer to this problem would be to allow communities to impose permit conditions; another would be to enhance the state's ultimate role in resolving community-developer siting disagreements, through some form of state override mechanism. Under this approach communities would not be allowed to reject a siting proposal outright but instead would have to take part in a state-administered process of binding arbitration. Communities would then have an incentive to undertake "good faith" negotiations because, in the absence of local cooperation, the state could authorize construction of hazardous waste facilities under terms that met state requirements but were less favorable to the community than if it had taken part in negotiations and wrested a "better deal" directly from the developer.[154]

The Massachusetts siting statute comes closest to this model, but with some important differences. Communities are allowed to reject a siting proposal outright, although this rejection requires a specific showing that the facility is more dangerous than other industries that are already accepted in the Commonwealth. In addition, a community does not have to take part in a state-administered process of binding arbitration, though drafters of the statute hope that the threat of binding arbitration would ensure a negotiated settlement without the arbitration. Also, it is not possible for any state agency to authorize construction of the facility. The state agency has to rely on the arbitration board to establish such terms, in the face of possibly sharp community opposition.[155]

A state siting board might best oversee the negotiations process. This is a common approach in those states that have already devised state siting policies, although this board normally would not exercise preemptive powers. Instead, it would facilitate agreements between the developer and the community. Ideally, the state agency would draw on its override powers only in the most limited of circumstances, trying instead to get the various negotiating parties to achieve a mutually satisfactory arrangement.

The Connecticut and Massachusetts statutes best embody this approach thus far, and the success of these two laws in the next several years will be a critical test for the negotiated approach to siting hazardous waste management facilities.[j] While it may be naive to believe that localities will ever be eager to negotiate with developers, use of state override powers in this eventuality is far superior to exercising preemptive power initially and thereby denying local interests a fair opportunity to participate in siting decisions.

Because of divergence in local interests, negotiations first must identify the various "interested parties."[156] Representatives of these parties must be explicitly included in the bargaining process for it to fulfill its goal of giving the various interests a stake in decisionmaking. At a minimum, this means including not only local officials as spokesman for the community but also abutters, community groups, and representatives of adjacent communities who might also be adversely impacted by the facility. The nature of the facility (that is, the specific design and technology employed) will dictate the extent to which spillover effects will be felt by the community and region and must be dealt with in the siting process.[157]

[j]In this regard, the siting problems in 1981 in Westford, Massachusetts, are illustrative and, to the proponents of a negotiated approach, somewhat worrisome.

A balanced siting strategy thus involves a mix of local authority, negotiations with the facility developer, and available (though tightly constrained) state override power. In order to negotiate effectively, the local community must have the power to compel the developer to do something. Without such authority, the local community doesn't really have the power to negotiate on behalf of its own legitimate interests.

At the same time, establishing state override in instances of local intransigence is part of a balanced siting approach. Decisions on sites for new facilities will almost surely prove unworkable in the absence of active state involvement, for local opponents would then have no reason to negotiate in good faith. Existence of such state override power does not necessarily make it usable by the state. Local rejection of certain facilities might well prevail, for use of this state power would be constrained by informal political pressures, formal mechanisms, and the threat of intransigent citizen opposition that always lurks in the background. Local opponents always can counter the state's override power with the effective forces of obstruction and disruption that they have at their disposal.[158] These are the political realities of a complex arena in which conflicting interests are at stake.

Nevertheless, reliance on negotiation with override—in contrast to preemptive approaches—allows local groups to have sufficient input into the final decision to safeguard their basic right to participate in decisions that affect the structure of their community, and thus build a modicum of respect for a legitimate decisionmaking process. For many people, local control over land use still symbolizes the ability of citizens to manage their own destiny. While some amount of state coercion probably cannot be avoided, much to the chagrin of those communities actually facing siting proposals, involving local interests as "equals" in the siting process through negotiation (backed by state arbitration) appears to be the most legitimate foundation for the exercise of this state coercive power.

The challenge to this approach comes from those who fear that successful, efficient decisionmaking is incompatible with high levels of citizen involvement or with a fair role for all of the siting interests.[159] The following comment in the report sponsored by EPA on public opposition to hazardous waste facility siting epitomized this viewpoint:

> . . . if given too large a substantive role, opponents will probably block all siting attempts. If major local roles in siting were to become widespread, this would probably be disasterous for siting.[160]

Perhaps the most important conclusion to be drawn from this analysis of alternative decisionmaking processes is that the conventional notions of public participation in siting should not be endorsed in new state siting policies. Successful siting should be judged not by the apparent ease with which decisions are rendered, but by whether these decisions account for the concerns of different individuals and groups and are seen by these different interests as fair. Although a process of this sort will not be easy to establish, it holds the best chance of long-range success. As William Singer, lobbyist for the New Jersey League for Conservation Legislation, warned the state Senate's Energy and Environment Committee during its consideration of siting legislation, "Shortcuts will not win public approval, and without general public consent, the proper siting of facilities will become impossible."[161]

NOTES

1. Robert Healy and John Rosenberg, *Land Use and the States*, 2nd ed. (Baltimore: Johns Hopkins University Press, 1979), p. 241.
2. Edmund Burke, *A Participatory Approach to Urban Planning* (New York: Human Sciences Press, 1979), p. 81.
3. W.R. Derrick Sewell and Susan Phillips, "Models for the Evaluation of Public Participation Programmes," *Natural Resources Journal* 19 (April 1979): 346.
4. Michael O'Hare, "'Not On My Block You Don't': Facility Siting and the Strategic Importance of Compensation," *Public Policy* 25 (Fall 1977): 411.
5. As an example of this problem, see Clark-McGlennon Associates, *A Decision Guide for Siting Acceptable Hazardous Waste Facilities in New England,* prepared for the New England Regional Commission (Boston, Mass.: November 1980), p. 43.
6. Based on O'Hare's "cast of characters," though changed considerably.
7. O'Hare, "'Not On My Block You Don't,'" p. 426.
8. See Chapter 2 and U.S. Environmental Protection Agency, Office of Water and Waste Management, *Siting of Hazardous Waste Management Facilities and Public Opposition,* SW-809, Report prepared by Centaur Associates (Washington, D.C.: U.S. Environmental Protection Agency, November 1979), p. 163.

9. O'Hare, "'Not On My Block You Don't,'" pp. 429–430.
10. U.S. Environmental Protection Agency, *Siting of Hazardous Waste Management Facilities and Public Opposition,* pp. 177–186.
11. Telephone interview with James McCoy, Stablex Corporation, 20 November 1981.
12. Fred Bosselman and David Callies, *The Quiet Revolution in Land-Use Control: Summary Report,* prepared for Council on Environmental Quality (Washington, D.C.: U.S. Government Printing Office, 1971).
13. U.S. EPA, *Hazardous Waste Facility Siting: A Critical Problem,* SW-865 (Washington, D.C.: U.S. Environmental Protection Agency, July 1980), p. 7.
14. Chemical Manufacturers Association, "A Statute for the Siting, Construction and Financing of Hazardous Waste Treatment, Disposal and Storage Facilities" (Washington, D.C.: CMA, 1980), pp. 18–19.
15. J. Gordon Arbuckle, "The Deepwater Port Act and Energy Facilities Siting: Hopeful Solution or Another Part of the Problem?" *Natural Resources Lawyer* 9 (1976): 512.
16. William Murray and Carl Seneker, "Industrial Siting: Allocating the Burden of Pollution," *Hastings Law Journal* 30 (November 1978): 318. Also see Michael Baram, *Environmental Law and the Siting of Facilities: Issues in Land Use and Coastal Zone Management* (Cambridge, Mass.: Ballinger, 1976), p. 144.
17. S.1300, 9 June 1980, sec. 16 (eminent domain), sec. 21 (preemption).
18. Booz, Allen & Hamilton, *Hazardous Waste Management Capacity Development in the State of New Jersey,* prepared for the State of New Jersey and the Delaware River Basin Commission (Bethesda, MD.: 15 April 1980), pp. IV–41, III–6.
19. Michigan, *Hazardous Waste Management Act,* Act 64, Effective 1 January 1980.
20. Letter from Governor William G. Milliken to Sally Cremeens, Princeton University, 26 August 1981, pp. 1, 4.
21. Jonathan Steeler, *A Legislator's Guide to Hazardous Waste Management,* prepared for National Conference of State Legislatures (Denver, Colo.: National Conference of State Legislatures, 15 October 1980), pp. 36–37.
22. National Governors Association, *Review of Sixteen State Siting Laws* (Washington, D.C.: NGA, 1980).
23. Alan Farkas, "Overcoming Public Opposition to the Establishment of New Hazardous Waste Disposal Sites," *Capital University Law Review* 9 (Spring 1980): 456.
24. Steeler, *A Legislator's Guide to Hazardous Waste Management,* p. 34.

25. See U.S. EPA, *Siting of Hazardous Waste Management Facilities and Public Opposition,* pp. 190–206.
26. Lawrence Bacow, *Mitigation, Compensation, Incentives, and Preemption,* prepared for the National Governors Association (10 November 1980), pp. 12–13 (emphasis added). Bacow presumably was not making a distinction between preemption and override.
27. Keystone Center, *Siting Non-radioactive Hazardous Waste Management Facilities: An Overview,* final report of the First Keystone Workshop on Managing Non-Radioactive Hazardous Wastes (Keystone, Colo.: Keystone Center, September 1980), p. 19.
28. Farkas, "Overcoming Public Opposition," p. 456.
29. S.C.S.1300, *Major Hazardous Waste Facilities Siting Act,* New Jersey Senate Committee Substitute, Senate approved 26 January 1981, sec. 11 (a).
30. William Murray and Carl Seneker, "Implementation of an Industrial Siting Plan," *Hastings Law Journal* 3 (May 1980): 1079.
31. Lester Thurow, *The Zero-Sum Society: Distribution and the Possibilities for Economic Change* (New York: Basic Books, 1980), p. 13. A rather different view is presented in David Morell and Grace Singer, eds., *Refining the Waterfront: Alternative Energy Facility Siting Policies for Urban Coastal Areas* (Cambridge, Mass.: Oelgeschlager, Gunn & Hain, 1980), pp. 181–256.
32. In California, as one example, the chemical industry and the hazardous waste management industry appear to have perceived this problem. After initially pressing for passage of a new preemptive state siting bill, industry adopted the stance that: "We prefer instead to operate only where we have the basic support of the local community. Otherwise we will face continuous opposition during our operations, and that's an impossible situation." (Personal communications to David Morell from several senior officials of the state of California and two industry representatives, November 1981.)
33. U.S. EPA, *Siting of Hazardous Waste Management Facilities and Public Opposition,* p. 310.
34. Bacow, *Mitigation, Compensation, Incentives and Preemption,* p. 13.
35. U.S. EPA, *Siting of Hazardous Waste Management Facilities and Public Opposition,* p. 307.
36. David Moldenhauer, "Toxic Waste Victims and Bureaucratic Accountability" (Term paper prepared for Engineering 303, Princeton University, 16 January 1981), p. 30. See Appendix A.
37. See Barry Casper and Paul David Wellstone, *Powerline: The First Battle of America's Energy War* (Amherst: University of Massachusetts Press, 1981).
38. Murray and Seneker, "Industrial Siting: Allocating the Burden of Pollution," p. 323.

39. The underlying philosophy of the Keystone 1 conferees, as suggested in their final report, seems somewhat at odds with this goal. The emphasis is on "involving a broad representation of the public and local governments," "educational programs," and "statesmanship." These actions are designed to "change" the "not-in-my-backyard syndrome" and thereby achieve "citizen approval of proposed facilities" (Keystone Center, *Siting Non-Radioactive Hazardous Waste Management Facilities,* p. 4). Education, diplomacy, and public involvement may all be to no avail, however, if the decisionmaking *process* is not seen by the citizens as sufficiently balanced.
40. Diane Graves, Conservation Chairman, New Jersey Sierra Club, Interview, 21 November 1980.
41. Edward Zukowski, in New Jersey Senate Energy and Environment Committee, "Public Hearing on S-1300: Volume I," 27 October 1981, p. 3A.
42. Hearing Record, State of California, *In the Matter of: Solving The Hazardous Waste Problem—Non-Toxic Solutions for the 1980s* (Los Angeles Convention Center, Los Angeles, California, 17 November 1980), p. 17.
43. David Moore, "Toxic Waste Authority Too Remote," *Windsor-Heights Herald* (June 26, 1980).
44. Barbara Hudson, New Jersey environmental activist, telephone interview, 13 March 1981.
45. David Deal, "The Durham Controversy: Energy Facility Siting and the Land Use Planning and Control Process," *Natural Resources Lawyer* 8 (1975): 451.
46. J. Douglas Peters, "Durham, New Hampshire: A Victory for Home Rule?" *Ecology Law Quarterly* 5 (1975): 67.
47. Connecticut Hazardous Waste Siting Board, *Final Report of the Interim Study Committee* (Hartford, January 1981), p. 8.
48. Florida, *Florida Resource Recovery and Management Act,* HB 311, Chapter 80-302, approved 2 July 1980.
49. Keystone Center, *Siting Non-Radioactive Hazardous Waste Management Facilities,* p. V.
50. Michael O'Hare, personal communication to David Morell, 19 January 1982.
51. See New Jersey's provision in Appendix B. New York's power plant siting statute provides $25,000 for the local community to commission its own studies as part of the state siting process.
52. Hearing Record, State of California, p. 16.
53. Casper and Wellstone, *Powerline,* pp. 135–136.
54. In the New Jersey facility siting case examined by Singer in "People and Petrochemicals, Siting Controversies on the Urban Waterfront," (in Morell and Singer, *Refining the Waterfront,* pp. 19–63), the state

environmental agency acted prior to a formal decision by the local government, causing greater anger at the local level.

55. The Chemical Manufacturers Association has devised a model state statute which includes a multimember state siting board. Similar institutions are now in place, or envisioned, in New York, Massachusetts, Michigan, Minnesota, New Jersey, and several other states.
56. O'Hare, personal communication, 19 January 1982.
57. Burke, *A Participatory Approach to Urban Planning,* p. 50.
58. Clark-McGlennon Associates, *Negotiating to Protect Your Interests: A Handbook on Siting Acceptable Hazardous Waste Facilities in New England,* prepared for New England Regional Commission (Boston, Mass.: November 1980), p. 37.
59. U.S. EPA, *Siting of Hazardous Waste Management Facilities and Public Opposition,* p. 14.
60. Ibid., p. 53.
61. McCoy, interview, 20 November 1981.
62. Singer, "People and Petrochemicals," pp. 42–43.
63. Ibid., p. 41.
64. Interview with Diane Graves and Katherine Montague, Princeton, N.J., 21 November 1980.
65. Interview with Joe McNamara, Port Authority of New York and New Jersey, 20 October 1981. Also see Madelyn Hoffman, "Ironbound Resistance to Waste Incinerator," *Re: Sources* (Winter 1981–82): 7.
66. David Bradford and Harold Feiveson, "Benefits and Costs, Winners and Losers," in Harold Feiveson, Frank Sinden, and Robert Socolow, eds., *Boundaries of Analysis* (Cambridge, Mass.: Ballinger, 1976).
67. Edward Haefele in *Representative Government and Environmental Management* (Baltimore: Johns Hopkins, 1973) argues that legislative institutions rather than executive agencies or the courts provide the best forum for making the complicated tradeoffs and balancing decisions inherent in environmental controversies.
68. The inherent imbalance between the utility calculus of a particular individual and of a political community is discussed in depth in Mancur Olson, *The Logic of Collective Action* (Cambridge, Mass.: Harvard University Press, 1971). The commitment of resources—time, money, or both—by the community may not be appropriate for the specific individuals.
69. Interview with Diane Graves, Princeton, New Jersey, 10 February 1982.
70. Richard Greenwood, "Energy Facility Siting in North Dakota," *North Dakota Law Review* 52 (Summer 1976): 723.
71. Murray and Seneker, "Industrial Siting: Allocating the Burden of Pollution," p. 335.

72. R. Kenneth Godwin and W. Bruce Shepard, "State Land Use Policies: Winners and Losers," *Environmental Law* 5 (Spring 1975): 715.
73. At the Senate Committee's meeting with environmental and public interest groups on 31 July 1980, for example, a Monroe Township resident and at least one other individual complained that advance notice of the meeting had been slight.
74. Diane Graves, Testimony to Senate Energy and Environment Committee, meeting with environmental and public interest groups on revision of S.1300, Trenton, New Jersey, 31 July 1980.
75. Godwin and Shepard, "State Land Use Policies," p. 712.
76. Graves, Interview, 19 March 1981.
77. U.S. EPA, *Hazardous Waste Facility Siting: A Critical Problem,* p. 4.
78. Diane Graves maintains that such citizens' interests *are* satisfactorily represented by the organized groups. Public participation, in this respect, simply means allowing all of these groups to enter fully into the decisionmaking process. Graves, Interview, 19 March 1981.
79. New Jersey Hazardous Waste Advisory Commission, *Report of the Hazardous Waste Advisory Commission to Governor Brendan Byrne* (Trenton, N.J.: State of New Jersey, 1980), p. 11.
80. John Skinner, U.S. EPA, Presentation to a national conference on "Groundwater in the '80's" (Chicago, 12 November 1981).
81. Michael Catania, Committee Staff to Senate Energy and Environment Committee at Workshop Session on S.1300, Trenton, N.J., 17 October 1980.
82. Based on a U.S. House of Representatives Subcommittee survey. See Chemical Manufacturers Association, "Overview of Hazardous Waste Problems," 29 January 1980. (Mimeo.)
83. See, for example, Monsanto's successful siting of a landfill in Bridgeport, New Jersey, that accepted only wastes produced by Monsanto, a company seen as "trustworthy and responsible." U.S. EPA, *Siting of Hazardous Waste Management Facilities and Public Opposition,* pp. 35–45. Also interview with McCoy.
84. Connecticut Hazardous Waste Siting Board, *Final Report of the Interim Study Committee,* p. 4.
85. Commonwealth of Massachusetts, *Massachusetts Hazardous Waste Facility Siting Act,* Chapter 21D, Approved 15 July 1980.
86. David Morell and Grace Singer, *State Legislatures and Energy Policy in the Northeast: Energy Facility Siting and Legislative Action* (Upton, N.Y.: Brookhaven National Laboratory, June 1977), pp. 199–229.
87. National Governors Association, Energy and Natural Resources Program, *Siting Hazardous Waste Facilities,* final report of the National Governors Association Subcommittee on the Environment (Washington, D.C.: National Governors Association, March 1981), p. 8.

88. Graves, Interview, 19 March 1981.
89. See, for example, Lester Milbrath and M.L. Goel, *Political Participation: How and Why Do People Get Involved in Politics?* (Chicago: Rand McNally College Publishing Co., 1977), p. 94.
90. Godwin and Shepard, "State Land Use Policies," p. 707.
91. Singer, "People and Petrochemicals."
92. Hoffman, "Ironbound Resistance to Waste Incinerator."
93. U.S. EPA, *Siting of Hazardous Waste Management Facilities and Public Opposition,* pp. 207-219.
94. J.F. Byrd, "An Industrial Approach to Siting of Hazardous Waste Disposal Facilities" (Speech to the National Conference on Management of Uncontrolled Hazardous Waste Sites, Washington, D.C., October 15-17, 1980), p. 16.
95. CMA, "A Statute," pp. 88-90.
96. Byrd, "An Industrial Approach to Siting," p. 16.
97. National Governors Association, *Siting Hazardous Waste Facilities,* p. 6.
98. O'Hare, personal communication, 19 January 1982. Professor O'Hare was closely involved in developing the NGA's position on facility siting.
99. Graves, Interview, 21 November 1980.
100. U.S. EPA, *Siting of Hazardous Waste Management Facilities and Public Opposition,* p. 9.
101. Ibid.
102. Sherry Arnstein, "A Ladder of Citizen Participation," in Godfrey Boyle, David Elliott, and Robin Roy, eds., *The Politics of Technology* (New York: Longman, 1977), p. 240.
103. Sewell and Phillips, "Models for the Evaluation of Public Participation Programmes," p. 352.
104. Arnstein, "A Ladder of Citizen Participation," p. 242.
105. This process was evident, for example, throughout the controversy over siting five energy facilities in New Jersey documented by Singer ("People and Petrochemicals"), pp. 19-63.
106. Arnstein, "A Ladder of Citizen Participation," p. 243.
107. Burke, *A Participatory Approach to Urban Planning,* p. 48.
108. S.1300, sec. 2 (b).
109. The public participation sections of the original S.1300 are found in sec. 11-13, 22-23.
110. Trafford, Testimony to Senate Energy and Environment Committee, meeting with county and municipal officials on revision of S.1300, Trenton, N.J., 11 August 1980.
111. The concept of "decide, announce, defend" has been articulated in Dennis Ducsik, *Electricity Planning and the Environment* (Cambridge, Mass.: Ballinger, forthcoming). Also see Ducsik and Don Shakow,

"Reaching Electric Facility Decisions: The Bankruptcy of the Decide-Announce-Defend Process," draft manuscript submitted for publication to the *Natural Resources Journal,* September 1981.

112. Milliken, letter to Cremeens.
113. Graves interview, 21 November 1980.
114. New Jersey's siting bill is supposed to ensure that hazardous waste facilities can be sited within the state. Hence, its critical component is the existence of an override provision—use of eminent domain authority—to locate these facilities despite anticipated local opposition. A second key provision of the bill is the establishment of a new institutional framework to accomplish this siting: the Hazardous Waste Facility Siting Commission. This body is independent and autonomous from the existing regulatory structure of the state's environmental agency (DEP). The commission's major functions are to: (1) devise a hazardous waste facilities plan which quantifies the need for additional waste facilities; (2) designate the appropriate sites to meet the needs indicated in this plan, by applying DEP-formulated siting criteria; and (3) condemn land for private operators who have obtained a DEP license and cannot otherwise obtain the site approvals required. The original bill allowed this independent commission (then a "corporation") also to construct and operate public facilities if the private sector had not succeeded in establishing adequate capacity. This provision was dropped after industry testified about its own ability to meet the need for an adequate number of facilities; in addition, industry objected to having these public facilities compete against private ones.

    The revised S.1300 has received widespread support from nearly all sides involved in the state's siting debate. The legislative development process addressed many of the discrepancies of the original bill. Most important, perhaps, were the greater assurances provided to local governments and the public under the revised bill. Instead of being able simply to attend a few public hearings on commission and DEP decisions, local citizens now are made part of these decisions by their membership on a Hazardous Waste Advisory Council which acts as advisor to both the commission and DEP on planning, siting, and licensing requests. Municipalities may contest those commission and DEP decisions that they view as unjustified. In addition, local officials in host communities will be trained to conduct their own inspections of these facilities. Finally, compensation will be paid to area municipalities to cover the increased costs of hosting a waste facility: police and fire protection, road repair, and so on.

    Increased public confidence in the process is also being sought by establishing within the bill provisions designed to mitigate certain environmental and health impacts. Notable in this respect are the limitations on construction of new "secure" landfills (to be allowed only as a means of last resort) and the adoption of certain basic siting criteria within the bill itself (for example, forbidding site selection near major sources of potable water).

Such measures have gone a long way toward increasing the apparent acceptability of the siting bill, which has addressed many of the concerns of the organized faction that has the least to gain from facility siting: local government. Industrial generators will be given places to handle their wastes. The hazardous waste industry is provided with a procedure for state assistance in locating its facilities. Finally, environmentalists, on the whole, recognize that construction of these facilities will provide safer disposal than illegal dumping.

115. Based on informal discussions with several leading hazardous waste experts in New Jersey, November 1980 (and thereafter).
116. Lawrence Susskind, *Citizen Participation in the Siting of Hazardous Waste Facilities: Options and Observations,* draft prepared for the National Governors Association (November 1980), p. 1.
117. Sewell and Phillips, "Models for the Evaluation of Public Policy Programmes," pp. 338–339.
118. Alvin Weinberg, "Science and Trans-Science," *Minerva* 10 (1972): 219–220.
119. Susskind, *Citizen Participation in the Siting of Hazardous Waste Facilities,* pp. 1–2.
120. Keystone Center, *Siting Non-Radioactive Hazardous Waste Management Facilities,* p. 15.
121. Weinberg, "Science and Trans-Science," p. 222.
122. Burke, *A Participatory Approach to Urban Planning,* p. 89. The three purposes of public participation used in this analysis are taken from Burke.
123. Keystone Center, *Siting Non-Radioactive Hazardous Waste Management Facilities,* p. 15.
124. Susskind, *Citizen Participation in the Siting of Hazardous Waste Facilities,* p. 2.
125. U.S. EPA, *Siting Hazardous Waste Management Facilities and Public Opposition,* p. 24.
126. Ibid., p. 60.
127. Ibid., p. 50.
128. Murray and Seneker, "Implementation of an Industrial Siting Plan," pp. 1070–1089. This is also true of energy facilities, including the Sohio Petroleum terminal and pipeline project in California which took five years before the developer withdrew his plans.
129. Tom Cook and James Knudson, "A History of Efforts to Acquire a Hazardous Waste Site in the State of Washington," Presented to the

Natural Resources Committee meeting of the National Conference of State Legislatures (Denver, Colo., 15–16 February 1980), p. 6.

130. Donald Mazziotti, "The Underlying Assumptions of Advocacy Planning: Pluralism and Reform," *Journal of the American Institute of Planners* 40 (January 1974): 44.

131. EPA's *Siting of Hazardous Waste Management Facilities and Public Opposition* is replete with examples of hearings which failed to produce adequate public participation. Also see Raphael Ebbin and Stephen Kasper, *Citizens and the Nuclear Power Controversy* (Cambridge, Mass.: MIT Press, 1975).

132. Murray and Seneker, "Implementation of an Industrial Siting Plan," p. 1085.

133. RCRA, for example, provides that public hearings must be held during the facility-licensing process. Douglas Costle and Eckhardt Beck, "Attack on Hazardous Waste: Turning Back the Toxic Tide," *Capital University Law Review* 9 (Spring 1980): 430.

134. Suggested by Greenwood, "Energy Facility Siting in North Dakota," pp. 726–727.

135. Diane Graves has described a session of this kind, held in Manalapan Township, New Jersey, one evening in November 1980. Present were local environmentalists, members of the local booster club, the Parents-Teachers Organization, and so on. A couple of them had heard Graves speak at a statewide League of Women Voters Forum on hazardous wastes, and invited her to Manalapan for a full-range yet informal discussion of the issues.

136. In general, local government hearings should be as open and informal as possible. The state needs to decide whether its own hearings are to be relatively informal, or structured and evidentiary (allowing cross examination of witnesses, and building a record suitable for judicial review under the state's Administrative Procedures Act). Who will chair the hearings? Where will they be located? (Hearings at night in the immediate vicinity of the site can have much greater impact than daytime hearings in the state capital.) How much prior notice will the public receive? Will appropriate documents be made available for citizens to review well before the hearing? Answers to all these procedural questions will, in total, communicate to the public the posture the state intends to adopt with respect to public participation.

137. U.S. EPA, *Siting of Hazardous Waste Management Facilities and Public Opposition,* p. 18.

138. Ibid., p. 42.

139. Advocates of nuclear power reportedly have spent a great deal of money to advertise their position in several state antinuclear public initiatives.

140. The 1976 California ballot initiative on nuclear power, for example, contained a great deal of very complicated wording. See W.C. Reynolds (ed.), *The California Nuclear Initiative: Analysis and Discussion of the Issues* (Stanford, Ca.: Institute for Energy Studies, Stanford University, April 1976).
141. O'Hare, personal communication, 19 January 1982.
142. The voters of Kern County, California, for example, were given an opportunity to express their opinions on a controversial nuclear power plant siting proposal (they voted "no"); and the voters of Atlantic County, New Jersey, voted two to one against the floating nuclear power plants proposed for a site off their county's shores. A Bordentown, New Jersey, referendum on a petrochemical facility is described in EPA's *Siting of Hazardous Waste Management Facilities and Public Opposition,* pp. 160–161.
143. A provision of this kind is included in the final version of New Jersey's S.1300.
144. Dorothy Nelkin and Michael Pollak, "The Antinuclear Movement in France," *Technology Review* 83 (November/December 1980): 37.
145. Irvin Bupp, "The French Nuclear Harvest: Abundant Energy or Bitter Fruit?," *Technology Review* 83 (November/December 1980): 32.
146. An example of this sentiment is the fact that much of the opposition to the original S.1300, especially from local officials, centered on the dangers of bestowing nearly all of the siting power in the hands of one autonomous agency.
147. Thurow, *The Zero-Sum Society,* pp. 11–19, 211–214; also see Godwin and Shepard, "State Land Use Policies," p. 707.
148. Clark-McGlennon Associates, *Negotiating to Protect Your Interests,* p. 11.
149. Booz, Allen & Hamilton, "Hazardous Waste Management Capacity Development in the State of New Jersey," p. IV–25.
150. Ibid., p. IV–26.
151. Clark-McGlennon Associates, *A Decision Guide for Siting Acceptable Hazardous Waste Facilities in New England,* prepared for the New England Regional Commission (Boston, Mass.: November 1980), p. 34.
152. Commonwealth of Massachusetts, *Massachusetts Hazardous Waste Facility Siting Act,* sec. 12.
153. Booz, Allen & Hamilton, *Hazardous Waste Management Capacity Development in the State of New Jersey,* p. IV–26.
154. The ability of this strategy to work rests on the fact that facilities are neither "safe" or "unsafe," but rather can be made more or less safe depending on the amount of resources expended in lessening spillovers and reducing risk. Weinberg, "Science and Trans-Science," p. 211.

155. O'Hare, personal communication, 19 January 1982.
156. Farkas, "Overcoming Public Opposition," p. 457.
157. This suggests that an agency like EPA should investigate the range of impacts associated with different types of facilities. These data could then be used to decide how far the radius of affected interests can plausibly be said to be; communities within this range should then be party to the siting decision. The limits of such a study, however, would stem from the problems of intangible or unquantifiable impacts being overlooked. Any "objective," quantifiable index of these impacts carries the danger of omitting important costs and benefits attendant on siting.
158. O'Hare, personal communication, 19 January 1982.
159. Sewell and Phillips, "Models for the Evaluation of Public Policy Programmes," p. 254.
160. U.S. EPA, *Siting of Hazardous Waste Management Facilities and Public Opposition,* p. 29.
161. N.J. Senate Energy and Environment Committee, *Public Hearing on S.1300: Volume I* (Trenton, N.J., 27 October 1980), p. 21.

5 WHO PAYS? COMPENSATION AND SITING

LEGITIMACY AND SELF-INTEREST

The preceding discussion has focused on the manner in which decisions on new facilities are made, particularly the issue of appropriate political authority. The recommended goal of creating a balanced, sequential, and timely siting process, including negotiation between local residents and the facility's developer, is to establish a system that is not only workable but acceptable to competing interests as a legitimate way to make these difficult decisions.

But is such legitimacy enough? The purpose of devising adequate decisionmaking mechanisms is to design a *process* that local citizens can come to respect, even though some of them may not accept a specific siting decision. With decisions as important as where hazardous waste facilities will be located, however, a legitimate decisionmaking system may be a necessary component of successful siting but is probably not sufficient by itself. If the citizens of Warrenton County, North Carolina, had been party to a legitimate, balanced siting process, for example, they probably still would have been vehemently opposed to the disposal of PCB's in their town. Attempting to obtain public acceptance of new hazardous waste management facilities based on creation of a satisfactory decisionmaking process ultimately involves asking residents near the site to rise above their own self-interest and assume some risks to achieve a broader collective benefit. This is asking a great deal of them.

Siting policies like that of New Jersey attempt to achieve satisfactory levels of public participation as a way to enhance public acceptance of these decisions. Such an approach essentially requires that an individual perceive it to be in the state's best interests for him to live near a hazardous waste facility (assuming that he is aware of the siting policy's net benefits overall). It also requires that the individual be willing to live near the waste facility, after he has been given an ample opportunity to register his own concerns and has been assured that the siting decision fairly takes account of his own interests as well.

This concept requires that individuals subjugate their own self-interest to a larger societal good. To expect this process to occur is not only "optimistic";[1] it may be fundamentally unfair as well. Hazardous waste facilities are generally a bad bargain for the residents of the host community. While the benefits of establishing a waste facility accrue to an entire state or region, the costs are borne primarily by the community where the facility is sited and by those adjacent to it. These costs result from a variety of adverse impacts associated with construction, operation, and maintenance of a facility: traffic, noise, air and water pollution, odors, aesthetic changes, risks, lowered property values, costly new public services, and community stigma. It is therefore *not* rational for local citizens to accept these facilities willingly, even on the basis of the most equitable and legitimate decisionmaking process.

"Mitigation," "compensation," and "incentives" are terms that have come to denote a range of ways to redistribute the costs and benefits of siting waste disposal facilities (and of other controversial large developments). These techniques may be particularly important in helping to reduce, though perhaps not to overcome, public opposition to siting. Such mechanisms can guarantee—albeit imperfectly—that those individuals who have to live near hazardous waste management facilities will not be forced to endure a situation in which they must suffer undue costs for the benefit of others.[a] According to this con-

[a]The argument here is philosophical. It is generally considered a legitimate enterprise for society to lower the utility of a minority within it so as to benefit the society as a whole. On the other hand, some limits clearly apply to such societal acts. It seems plausible, for example, to maintain that the decrease in the utility of each minority individual ought not to be significantly larger than the increased utility of each majority individual, unless the minority is "paid back" in some way. In other words, the costs and benefits that result from a policy should be redistributed so that the per capita losses do not greatly exceed the per capita gains. This notion could be used to justify "takings" law, in which the government must compensate landowners for any property that was seized by the state. Not doing so would, by the argument advanced here, be "unfair."

ception of minority rights, political coercion becomes legitimate when the victims of an action are not only parties to the decision-making process but also are compensated fairly for the substantial harms they may suffer.[2] The Booz, Allen & Hamilton report on siting in New Jersey maintains that ". . . communities have a legitimate right to expect to be compensated for being willing to accept the costs of these facilities."[3] Conceptually, this is not bribery for accepting the facility, but fair compensation for localized harm.

QUESTIONING THE TRADITIONAL APPROACH

The implications of this compensation argument are far-reaching and controversial. Clearly, the idea of accompanying a policy that imposes considerable costs with means to offset these costs goes against the traditional tendency to "let the burdens fall where they may." Many public policies, of course, have redistributive effects. By imposing costs on some and bestowing benefits on others, such policies change the relative distribution of wealth. Nonetheless, policymakers in the United States have not embraced a comprehensive ethic of compensating the losers of a given policy. Instead they appear to be willing to allow these redistributive effects to occur. The principal exception to this precept lies in the concept of a "taking," a term used to describe "publicly inflicted private injuries" for which compensation is constitutionally required. Takings are distinguished from "valid exercises of state police power," and although the latter may also have the result of making some people worse off, such actions do not require compensation.[4] Although the line between these two is far from clear, an obvious example of a compensable taking occurs when a state seizes private land for use in a particular public project. The state in this case pays the land owner fair market value for his property. Land-use regulations, on the other hand, have generally been deemed by courts to be outside the class of takings[5] and thus not subject to constitutionally mandated compensation.[6]

The reluctance to compensate individuals for the costs they receive from a given policy is potentially well grounded. Inasmuch as compensation depends on redistributing the net benefits from a particular policy to "pay back" the losers, the potential gains from a policy are lessened by the costs of establishing compensatory mechanisms. Indeed, in some cases the costs of compensating individuals harmed by a policy might exceed the benefits that would be realized in the absence of compensation, so that the "settlement machinery eats up the gains."[7]

There is another reason not to compensate the losses that accompany each policy: the belief that public policy, overall, eventually evens out the distribution of costs and benefits. By this view, today's loser in a siting decision is tomorrow's winner in some other redistributive governmental action. This "logrolling" notion, according to Frank Michelman, is based on the premise that

> . . . even though particular measures cannot be shorn of capriciously redistributive consequences, we can arrive somehow at an acceptable level of assurance that over the life of a society (and within the expectable lives of any of its members) burdens and benefits will cancel out leaving something over for everyone, and that society ought, therefore, to proceed to economize its resources, using governmental coercion where necessary and not agonizing too much over compensation.[8]

This justification for eventual net redistribution is not particularly convincing, however, in the case of hazardous waste facility siting. The burdens imposed locally by these facilities are frequently perceived as being much higher than those associated with virtually any other contemporary public policy. As Malcolm Getz and Benjamin Walter have pointed out, "Almost any other conceivable use of the land is preferred to a hazardous waste dumpsite."[9] In New York, for example, a proposed hazardous waste facility has encountered much more intense local opposition than did a nuclear power plant proposed earlier for the same community. Hundreds of redistributive policies in which a community is on the receiving end of the benefits might be required to offset the losses bestowed by a single hazardous waste siting act.

Even with their high localized costs, if a state needed many different sites, the logrolling approach might still be persuasive. At its extreme, imagine the siting of a waste facility in every municipality, thereby effectively evening out the losses imposed.[b] Once again, however, this is clearly not the case with siting hazardous waste facilities.

[b]There are potential problems even with this hypothetical set of circumstances, however. Most obviously, there are many different types of hazardous waste management facilities, some of which impose greater costs on a locality than do others. With these differing costs, communities would attempt—in the absence of compensation—to obtain those facilities that were least costly and to avoid those with the greatest costs.

In New Jersey, for example, state officials project a need to site as few as five new facilities over the next several years.[10] Similar numbers probably apply in California.[11] In the absence of procedures for fair compensation, the costs of siting hazardous waste facilities will necessarily be concentrated within only a very few communities.

The traditional noncompensatory approach to land-use regulation and control frequently ignores the redistributive aspects of siting. Since many of the costs associated with hazardous waste facility siting involve quality-of-life impacts, these costs are not prone to quantification. Even if greater use of compensation were attempted, difficult problems would arise in determining how to offset these unquantified costs. Here developer-community negotiation processes become especially important.

Moreover, certain individuals and groups are more likely than others to wind up on the losing end of decisionmaking. These people, in general, lack the political and economic resources to claim their share of the policymaking pie. R. Kenneth Godwin and W. Bruce Shepard point out:

> The persons currently excluded from the politics of statewide land use planning are the poor. This is unfortunate, in that they are obviously affected by the consequences of limitations on the amount of land for development of new housing or the siting of public and private facilities.[12]

Decisionmaking processes tend to favor organized interests—those with a strong structural dimension, as shown in Chapter 3—to the detriment of groups that lack political power. While some aberrations in this principle have begun to appear, it is still typical of "nuisance" facility siting that, "The stronger politically influential communities tend to deflect such installations in the direction of their weaker neighbors."[13] Political strength in these instances serves to reinforce economic and social advantages—and vice versa.[14]

SITING STRATEGIES AND THE DISTRIBUTION OF COSTS AND BENEFITS

The way in which a siting policy allocates costs and benefits is central to its success. As noted already, siting concentrates high costs in the community, while providing diffuse statewide benefits. The National Governors Association has stated quite bluntly: "We cannot assure citizens who live near a proposed site that the landfill, incinerator, or treatment plant will have no [adverse] effect on the community."[15]

Success in site approval, while obviously important, should not be the controlling principle. Selecting winners and losers in siting also involves important issues of equity and social justice; and perceptions of these issues ultimately will determine the fundamental legitimacy of the siting process. As David Ervin and James Fitch have pointed out:

> Even assuming that the benefits of a proposed land-use planning action outweigh the costs, this does not guarantee that the particular land-use allocation is desirable from a general social point of view. The question of who receives the benefits and who bears the costs is also vitally important, for it is this pattern of distribution which determines whether or not the planning measure meets social norms for equity or fairness.[16]

This concern for the wider social implications of siting is all too frequently overlooked in discussions of alternate approaches, particularly those which stress siting success and use of preemptive authority to gain site approval.

The Path of Least Resistance: Land-Use Compatibility

Much of the literature on siting hazardous waste management facilities seems to agree that locating these facilities is easiest in industrialized areas. In the absence of state preemptive authority over siting, facility developers have generally been more successful in these locales. The Centaur Associates study sponsored by EPA on public opposition to siting hazardous waste facilities examined twenty-one cases where sites had been proposed; the four in which there had been little or no opposition were all located in industrialized regions.[17]

Many conclude from this evidence that successful siting policies must concentrate on establishing hazardous waste facilities in industrial areas where opposition is often minimal or nonexistent. Alan Farkas has suggested that ". . . in identifying sites, agencies should focus on those areas which are most likely to be publicly acceptable." These, in his opinion, are in industrialized areas "where nearby residents have become accustomed to the nuisances and hazards associated with industrial operations."[18]

Industrial areas appeal to those responsible for selecting sites for new hazardous waste facilities for other reasons as well. Here local zoning regulations would not pose a serious obstacle. Such facilities would be compatible with existing land uses, thereby minimizing the facility's overall social costs. Wastes would not have to be moved long distances from the industries which generated them to the facilities for their treatment, storage, and disposal. Moreover, many industrialized areas simply do not have very many people living in the immediate proximity of the industrial facilities. While many people live in New Jersey, for example, making it the most densely populated state in the country, in this state's industrial areas relatively few houses are actually interspersed among the industrial facilities.[19]

From a political perspective, the strategy of siting facilities in industrial areas where opposition is weakest might be termed the "path of least resistance" approach. Limited opposition to siting in industrial areas, however, should not be attributed primarily to greater tolerance that residents of these areas have for the hazards and pollution associated with hazardous waste facilities. Rather, a crucial factor in this lack of opposition is the fact that residents of industrialized areas are less likely than residents of other areas to be able to muster the political resources needed to oppose siting attempts effectively. Industrialized areas are typically within an urban community surrounded by predominately lower and lower middle-class residents. These individuals may be susceptible to the traditional practice of imposing burdens on those who are least able to oppose them; normally these groups are not organized and lack political power. This particular scenario is confirmed by Robert Healy and John Rosenberg who state: "Often it is the city with the least affluent, least articulate electorate that winds up with the facilities rejected by other jurisdictions in the region."[20]

To a far greater extent than in the past, however, groups in urban areas now appear to be attempting to keep noxious facilities away. Citizens along New Jersey's Hudson River waterfront, for example, defeated five energy and petrochemical facility-siting proposals in a row in the 1970s.[21] In Elizabeth, New Jersey, a community group known as the Coalition for a United Elizabeth (CUE) has taken up the issue of hazardous wastes. As Sister Jacinta Fernandes, Associate Director of CUE, testified at one of the hearings on S.1300, "We in Elizabeth have had it. We have been literally dumped on."[22] Residents of Newark's Ironbound community have organized to oppose a new waste facility:

> Ironbound's diverse population is no longer willing to let its neighborhood become the toxic waste dumping ground for New Jersey. . . . Over a year ago, residents formed the Ironbound Committee Against Toxic Wastes with two demands: clean up all toxic waste sites already in Ironbound and prevent any more toxic wastes from coming into the neighborhood.[23]

Groups of urban residents increasingly see themselves as having been victimized by public and private actions that have imposed great burdens on them. While for years such people seemed—by their silence—to have become "accustomed" to these costs, these people are now beginning to erect obstacles to the path of least resistance. Such opposition is founded on notions of political equity. Increasingly, those who live in urban-industrial areas are articulating their resistance to the systematic use of their areas for facilities that endanger the environment and public health. As one activist opposing the siting of a petrochemical facility in Jersey City, New Jersey, asked: "Are we second or third class people condemned to have, see, smell, breathe and slosh through nothing but oil and its products?"[24] Sister Fernandes told the Senate Committee, "I would say the majority of us have suffered some type of health disorder which may very well be attributable to the fact that we are in the heart of the petrochemical industry."[25]

Siting policies like that of the National Governors Association,[26] which explicitly endorse a path of least resistance through emphasis on locating these facilities in industrial areas, appear to be unsatisfactory in at least two major respects. On the one hand, the emergence of greater citizen activism in urban areas threatens to produce opposition to proposed facilities in places where it was unlikely to be found in the past. As Ronald Luke concluded:

> Today, industry cannot indulge in the luxury of siting its facilities or developing natural resources in areas where there is no risk of community rejection. Such areas probably are nonexistent. In every situation, there is some threshold of impacts or perceived impacts at which the siting of a facility will be seriously questioned or opposed.[27]

Hazardous waste facilities definitely have surpassed this "threshold" and have begun to generate opposition even among groups that otherwise would be unlikely to organize.

Moreover, even if locating hazardous waste management facilities in industrial areas continued to correspond to a traditional pattern of minimal opposition, singling out these areas to shoulder the burdens of waste disposal expands on present injustices. The path of least resistance runs directly counter to the goal of equal treatment of citizens. If they are subjected to the dangers of hazardous substances by policies that restrict siting to industrialized areas, residents of these areas are quite justified in their claim of being treated like "second or third class people." In sharp contrast to industrial-area siting, Healy and Rosenberg advocate the more equitable view that "policies should consider the interests and aspirations of all users." Thus, when government becomes involved in resource allocation, they say, it must not "unduly favor certain income groups or certain types of use."[28]

In sum, siting of hazardous waste management facilities in urban-industrial locales may offer a number of net advantages, at least in comparison with many residential or rural locations. However, it is certainly no panacea for the developer seeking community silence on urban-industrial sites.

The "Fair Share" Debate and Political Equity

The premise of the foregoing argument is that the burdens of hosting a hazardous waste facility ought not, as a matter of policy, to be restricted to industrialized communities. Such a policy would unfairly single out residents of these areas to bear the net social costs of hazardous waste disposal. Whereas nonindustrial areas also share in the benefits of industrialization (for instance, consumers throughout the state purchase products made in waste-generating processes), it is unfair that they should recieve these benefits without paying at least some of the attendant disposal costs.

Residents of nonindustrial areas, however, are likely to turn this equity argument around with equal conviction, arguing that restricting the costs of siting to industrial areas is fair since these are the regions that derive most of the direct economic benefits from industries that generate waste (for example, tax payments and jobs). From their perspective, locating a hazardous waste facility in a nonindustrial community unfairly provides this area with most of the costs and all too few of the benefits connected with toxic substances. The report to EPA on public opposition found that:

> Residents of rural areas have expressed opposition to accepting wastes generated by urban industries. . . . Their objections are based on the likelihood that they would be bearing risks associated with these wastes while others receive the benefits.[29]

Fears of nonurban residents concerning the heavy toll that would be exacted by locating a hazardous waste facility in an area zoned for nonindustrial use led Mayor John McCarroll of Washington Township, New Jersey, to propose that his state's siting bill incorporate a provision requiring treatment facilities for hazardous wastes to be located within a fifteen-mile radius of the generators of such waste.[30]

This equity argument against siting hazardous waste facilities in nonindustrial areas is buttressed by the appeal to land-use compatibility. To the extent that hazardous waste facilities "fit in" with similar existing land uses, they impose fewer costs on industrial regions. This type of compatibility was exemplified by Frontier Chemical Waste Processors' success in locating a treatment facility on the site of a former chemical plant in Niagara Falls, New York. This facility's operations were relatively inconspicuous, consistent with previous activities at the site, and conformed to adjacent land uses.[31] In such a case, the local costs were probably not as high as they would be for nonindustrial siting since: ". . . the perceived impact of a hazardous waste facility on the environment and/or quality of life in an area is likely to be greater where similar operations do not already exist."[32]

Opposing views about the fairness of where facilities ought to be sited are difficult to resolve. Although urban residents seem justified in saying, "Hey, you guys, it's your turn to take the risks,"[33] nonindustrial areas can rightly claim that they are being asked to suffer tremendous costs while other regions gain all the benefits. Both seem to be legitimate appeals to equity, bearing out Frank Michelman's contention that ". . . fairness resists being cast into a simple, impersonal, easily stated formula."[34] Despite the apparent irreconcilability of these two viewpoints, both share a common perception of hazardous waste facilities as a bad bargain whose impacts on both industrial and nonindustrial areas are generally *not* offset by sufficient benefits to be acceptable to local residents, whether urban or rural.

In a few instances, though, siting efforts have successfully tied the need for waste disposal to the benefits associated with processes that create waste. The Gulf Coast Waste Disposal Authority, for example, successfully sited a landfill in Texas City, Texas, within a highly industrial area. According to the EPA-sponsored study of this case, "The local residents appreciate the fact that [local] industries must dispose of wastes and that a disposal site in their area is just one more fact of life."[35] Likewise, public acceptance of on-site industrial disposal of wastes is likely to be higher than for commercial (off-site) facilities since an evident relationship exists between the generating company and its need to dispose of its wastes. In the case of the Monsanto landfill in Bridgeport, New Jersey, because the generator was a key actor in the community, residents did not oppose the construction of an on-site facility. At the public hearing, Monsanto noted that "the landfill was needed if manufacturing were to continue." Thus, as the EPA-sponsored report indicates, there was a clearly perceived link between the provision of community jobs by the generator and its need for disposal capacity.[36].

To conclude that industrial areas should be fully responsible for disposal of all their own wastes does not, however, seem justified. In the case of the Gulf Coast Authority, for example, public participation was strategically bypassed to avoid public opposition. Had this opposition been given an opportunity to emerge, many residents might well have expressed views similar to CUE's or the groups in Ironbound or Jersey City. The waste-generating industries, while providing tangible benefits, also are responsible for so many costs that the added burden of hazardous waste disposal is likely to be perceived by many in the urban community as an unfair net additional cost not balanced by sufficient additional benefits.

SITING CRITERIA: HIDDEN DEFECTS

The conflict over which areas ought to accept their "fair share" by hosting hazardous waste facilities holds the power to polarize political debate along an urban and rural dimension. In New Jersey, this conflict may well be played out as an issue of north versus south, particularly as politicians from the southern half of the state attempt to dramatize the evils of hosting facilities for wastes generated in the northern part of the state.[37] Similar tensions could emerge in California, Illinois, Michigan, and other states with large urban-industrial concentrations and also large rural areas.

One way to try to resolve these conflicts over where to site facilities is to establish explicit siting criteria which identify suitable locations within the state for such facilities.[c] Determining appropriate siting criteria, however, entails a more complex process

[c]See the discussion in Chapter 4 of the distinction between using exclusionary criteria and using criteria as a checklist to guide political decisions.

than merely rendering a decision on a developer's proposal to construct a facility within a given community. Siting criteria and site inventories allow the state to use its own initiative, prior to receipt of developers' applications. The usual idea is to identify the "best" locations in the state—or at least a certain number of adequate locations for these needed facilities. Though environmental and health impacts of hazardous waste facility siting cannot be eliminated altogether, the adverse consequences of such facilities might be reduced through careful site selection. Although complete safety measures can never be guaranteed (that is, there is always risk), experts argue that the consequences of any potentially deleterious impacts should be reduced by placing the facility in a location which meets certain technical criteria for "adequacy." Moreover, the very process of openly ranking several different sites against specific criteria might help increase the credibility of the decisionmaking process. Development of siting criteria thus offers a unique opportunity to blend technical expertise with political sensitivity.

Doing an inventory of suitable sites in the state could have an important strategic component, shifting critical initiative from private to public hands. Traditional methods of site selection by a developer mean that ". . . local citizens can justifiably argue that a private firm's selection of a site in their region is arbitrary and unfair."[38] In contrast, a state inventory of appropriate sites would allow a developer—who then locates on a site deemed suitable—to respond readily to a community's challenge that "better sites exist elsewhere." Many local residents, no doubt, still will be unhappy, still will be opposed to the location of the facility; but the existence of explicit siting criteria will make it somewhat more difficult for them to ask, "Why not locate it somewhere else?"

Using an approach of this kind, the Minnesota Waste Management Board in 1981 proposed eighteen areas of the state for possible inclusion in a formal list of preferred locations for hazardous waste facilities. The board announced plans to promulgate a final inventory, including at least three possible sites each for incinerators, chemical processing facilities, and transfer and storage stations. This inventory was intended as "a way of encouraging industry to site facilities that provide alternatives to land disposal."[39] The board has also proposed twenty areas as possible sites for a new commercial hazardous waste landfill. Although Minnesota's 1980 Waste Management Act does not give the board authority to operate waste facilities nor to issue siting permits, this agency does have the power to override local vetoes of attempts to site hazardous waste facilities, provided that the proposed site is located in one of its designated "preferred areas" and that the facility operator has obtained necessary technical and environmental permits from the state's Pollution Control Agency. The Waste Management Act requires the board to designate at least one site as suitable for a commercial landfill, and to include such a facility in its final waste management plan. Minnesota found, however, that its proposal for eighteen areas as formal locations for waste facilities generated so much controversy that the board reversed course completely and interpreted their statute to allow them to designate the entire state as a potential site location area.[40]

Deciding what constitutes an optimal site for a hazardous waste facility is not easily accomplished. The process involves using scientific and technical procedures to identify the characteristics of a site and thereby to determine its suitability or nonsuitability. Indeed, a critical ambiguity often surrounds the siting criteria concept. Traditionally, such criteria have been seen as a series of absolute exclusionary rules which forbid the siting of facilities in certain areas.[41] That is, criteria normally would prohibit landfills and similar facilities from locating over a major aquifer, since an accident there could contaminate drinking water. Sites near large numbers of existing residences might also be excluded. Conversely, other factors indicate an area's potential suitability; for example, the existence of a dense layer of clay might provide some natural protection against the spread of certain contaminants.

A large number of factors are potentially relevant to determining a site's suitability. These factors include not only information on soil permeability and groundwater quality but also whether the site is near historic places or close to "incompatible structures." In essence, some criteria are of almost unquestionable importance—flood hazard areas, geologic faults, public water supplies, wetlands; many others vary in importance. Inevitably, since no site is perfect, tradeoffs must be made between competing value systems. For example, is it better to locate a facility near a community of moderate population density, or to have it jeopardize critical, though remote, habitats for rare and endangered species?

An alternative to the exclusionary approach has therefore been suggested. This view holds that siting criteria should be used flexibly, as a guide to the kinds of issues that the state licensing authority or the local governing body should consider. The experience in both New Jersey and Connecticut in using criteria as exclusionary rules has been distinctly unpromising. Once people perceive that their own community has not been excluded as a possible site, the earlier consensus over acceptable criteria has rapidly dissolved into opposition to the proposed facility, along with demands that the siting criteria themselves be altered.[42]

Some states may find it useful to devise systems whereby local governments—perhaps counties—adopt siting criteria developed by the state in their own local ordinances, and then make decisions based on these criteria. This mode of decisionmaking has been used in a number of cases, often with good results: among others, land-use control in wetlands and floodplains, and California's approach to land development decisions in its coastal zone.[43] Local decisions of this kind might enhance public acceptance of siting hazardous waste facilities. Use of siting criteria to select sites that are most environmentally compatible with hazardous waste facilities differs greatly from the path of least resistance approach, since the latter employs "siting methods that are essentially opportunistic, in that the methods of site selection are designed . . . irrespective of the environmental consequences."[44]

It is neither legitimate nor wise, however, to endow siting criteria with any more validity than they deserve, although unfortunately, this is often the case. People refer to the criteria as objective measures—scientific justifications for preferring one site over another. The New Jersey Governor's Hazardous Waste Advisory Commission, for example, referred to the need to devise "objective selection criteria" prior to actual selection of sites.[45]

The process of developing siting criteria is itself an inherently political one. Though the criteria rely on many technical measurements and comparisons, they are not without their own significant political components, particularly when requiring tradeoffs in potential damages of various kinds or in deciding how heavily to weight any particular factor (proximity to existing residences, for example). In the process of developing statewide criteria, people may examine their own community's characteristics in order to try thereby to exclude their own area from the list of potential candidates.[d]

[d]In New Jersey, for example, East Windsor Township—angered at being considered the possible location of a hazardous waste site—has proposed extending the population proximity criterion from 500 to 1,000 yards, thereby excluding that particular municipality from further consideration. An earlier controversy in Bordentown, New Jersey, led to similar proposals. In this case, the area's state assemblyman introduced a bill calling for a ban on any hazardous waste landfill within 1,000 feet of a school, effectively prohibiting the facility as proposed.[46]

Thus several major drawbacks affect use of siting criteria, particularly exclusionary ones. Some sites may score superbly in every dimension except one. If the criteria are indeed absolute, such a basically acceptable site would have to be excluded from further consideration. Instead, the decisionmaking body would have to opt for a site which only barely satisfied each of the criteria.[47]

Moreover, despite laudable intentions of putting hazardous waste facilities in the "best" sites possible, arriving at a consensus over acceptable criteria may well be very difficult. It may also prove impossible to make that consensus hold after it has been achieved. Those communities that are involved will bring different values to the process, values corresponding to their own particular situation. Nonurban residents may demand that sites be put in places where they are fully compatible with existing land uses (siting in urban-industrial areas); urban residents, in turn, are more likely to favor criteria that exclude sites near large population centers (siting in rural areas). The debate over siting might thus simply be shifted forward to the debate over siting criteria since the outcome of decisions on criteria would set the direction for future siting. The Delaware River Basin Commission (DRBC), jointly with the state's Department of Environmental Protection (DEP), for example, established tentative criteria for siting hazardous waste facilities in New Jersey (and elsewhere in the Delaware River Basin). These DRBC-DEP criteria emphasized a site's geological characteristics (for instance, aquifer use, groundwater flow, soil permeability).[48] When made public, the reaction from communities not in the excluded areas—East Windsor Township, New Jersey, for example—was intensely negative.[e]

These types of conflicts might be satisfactorily worked out by means of a negotiated or mediated decisionmaking process, were it not for the unlikelihood of all of the relevant siting interests being brought into this process. A crucial drawback to the creation of siting criteria is that because they are decided upon before actual facility proposals are submitted, their determination lacks the "salience" of actual decisions over whether to site a particular hazardous waste facility. The National Governors Association has cautioned that site inventory processes rarely gain public interest until sites are identified, after which the criteria themselves are challenged.[50] Consequently,

[e]This may not have been a fair test for the siting criteria concept, however. In this case, the map was poorly drawn and the procedure was badly explained to local residents, thus accentuating the negative reaction.[49]

even if a consensus is reached on siting criteria, it is likely to be short-lived since groups that did not participate in this process later reject the criteria when they become involved in actual decisions on sites. Thus, although the concept of siting criteria is compelling, the possibilities for arriving at criteria through a legitimate decisionmaking process seem, from a practical standpoint, extremely remote. To the extent that these criteria might be challenged anew during each siting attempt, their use might even prove a hindrance in the process of selecting suitable locations for hazardous waste management facilities.[f]

REALIGNING COSTS AND BENEFITS

Mitigation: Cost-Reduction Through Facility Improvements

The assumption behind use of siting criteria is that facilities should be located where the damage they might inflict will be less severe than if they were sited elsewhere in the state. This approach recognizes that hosting a hazardous waste facility may impose substantial costs.[51] The siting criteria identify sites whose characteristics are such that these costs will at least be minimized.

Regardless of whether siting criteria are themselves an effective means of achieving this aim, cost minimization is generally an integral part of all facility-siting efforts. Regulations for construction and operation of different types of facilities can reduce their harmful effects. The regulatory approach lays out minimal protections against spillover effects from facilities. The objective of such regulations is to make facilities adequately "safe" by preventing or reducing their impacts. This action is referred to as "mitigation":

> Mitigation represents the first line of defense in addressing local concerns and will reduce the need to provide compensation and incentives. Mitigation may involve, for example, redesigning a facility to provide extra protection against groundwater pollution or noise.[52]

[f]The negotiated mode of decisionmaking is unfortunately not appropriate to the formulation of siting criteria. It would be necessary to include *all* of the municipalities of the state to have a truly representative decision, but this would be an unwieldy number of actors. This would also be no guarantee of including the many nonstructural groups with a big stake in siting. This argument is not intended to reject the use of siting criteria but merely to point out some limitations. At this point, their use in processes like New Jersey's does not portend success.

The rationale for use of mitigation, according to a report prepared by the U.S. Environmental Protection Agency, has four components. First, it is usually far more efficient to deal with impacts up-front, rather than paying for them later. This proposition is borne out by the fact that although over $30 million has already been spent trying to deal with contamination of the Love Canal in Niagara Falls, New York, proper disposal of these Hooker Chemical Company wastes in 1952 would have cost only $4 million (in 1979 dollars).[53] The additional three reasons for mitigation are:

- Minimizing impacts will better enable the hazardous waste facility to develop a "good neighbor" track record—which is especially important for staying in business.
- Mitigation minimizes the problem of estimating the cost of impacts and negotiating compensation.
- Mitigation may be strategically useful in demonstrating commitment and credibility when negotiating with local groups.[54]

Use of mitigation, then, may well account for the success of a particular siting proposal inasmuch as these measures reduce the ratio of costs to benefits for a community. The drawback to mitigation measures, however, is that these changes in facility design and operation generally address only the *tangible* impacts associated with hazardous waste facilities. Many measures involve changes in facility design and operation: for example, additional or thicker liners for landfills, a better stack gas scrubber for an incinerator, changes to an access road, incorporation of a buffer zone around a facility, additional monitoring, independent inspection agreements, or institution of extra safety and security measures through purchase of extra equipment or employee training in emergency response techniques. Design changes are most suited for reducing tangible impacts such as traffic, noise, aesthetics, or air and water pollution. Intangible impacts such as risk, community stigma, and property value changes are less subject to design mitigation. Design changes can lessen these problems, but not eliminate them.[55] The mitigative strategy, therefore, is quite consistent with conventional approaches to siting. Although costs to a community are reduced, mitigation by itself still allows these costs to "fall where they may."

Existing institutions and siting procedures are basically able to accommodate mitigation. Prior to the approval of new facilities, for example, an Environmental Impact Statement (EIS) or similar state-mandated assessment will consider appropriate mitigative measures which might be added to the design of a proposed facility. In the case of a hazardous waste incinerator, for example, this might mean the addition of a scrubber to the incineration stack.[56] Existing use of error-detection strategies are likewise potentially useful vehicles to identify possible mitigative changes in a proposal. Thus, hearings before a state environmental agency on a license application offer an opportunity to engage in these design changes, particularly if these measures are made a condition for facility licensing.[57]

To the extent that mitigation still leaves the balance of costs and benefits to an area skewed toward the costs, its strategic success will be limited. Nevertheless, mitigation can certainly contribute to an overall siting strategy. Its weaknesses will be most profound when individuals perceive high intangible costs associated with a hazardous waste facility. Even with the best mitigative devices, these costs will remain—and so, in all likelihood, will much of the public opposition.

Compensation: The Redistributive Approach

The imbalance between costs and benefits can be remedied more completely through use of compensation. In theory, the purpose of compensation is to make affected communities *as well off* after the siting event as they were before. Compensation is, therefore, more far-reaching than mitigation because it is designed to deal with unavoidable, intangible, or uncertain impacts. Compensatory techniques generally include monetary payments, in-kind replacement of resources or services, and provision for contingency funds. Monetary payments can take numerous forms, including property taxes, payments in lieu of taxes, tipping fees, gross receipts taxes, special impact payments, state-local aid adjustments, and lump sum payments.[58] The innovative monetary compensation schemes instituted in Connecticut and Kentucky exemplify this process. In Kentucky, the facility pays a 2 percent gross receipts tax to the host county; in Connecticut, the operator pays a tipping fee of five cents per gallon plus a gross receipts tax which varies from 2.5 to 10 percent depending on the facility's overall level of revenue. Monetary payments are best suited for dealing with those impacts for which costs are readily definable. Other issues such as a tarnished community image (for example, becoming a state's "dumping ground") or a facility's risk to public health are less appropriate targets for monetary compensation, though even here compensation can be a crucial component of an overall siting strategy.

The potential importance of compensation within a siting process is suggested by one individual's letter to the editor of a Seattle newspaper:

> Coating such bitter pills [nuisance facilities] with a bit of sugar would undoubtedly remove much opposition. It is when the undesirable is jammed down the throat without compensation that people feel unfairly treated.[59]

The report to EPA on public opposition to siting confirms this problem: "Opponents question the fairness of having their town bear such a large share of the environmental costs of modern industry."[60] Instead of focusing on the fairness of the siting process itself, one goal of compensation is to provide a more equitable outcome.

Compensation may also play a strategic role by making hazardous waste facilities "easier to swallow." This use of compensation assumes that individuals will not accept hazardous waste facilities unless it is in their own interest to do so. Michael O'Hare contends that:

> The siting problem characterized by "not on my block you don't" is a consequence of the basic strategic configuration of the problem; it will not respond to improving the different actors' knowledge about each others' interests. . . . Compensation for local sufferers is not only an equitable desideratum, as has long been recognized, but a strategic necessity for aligning critical actors' interests with the public interest.[61]

The underlying assumption of compensation is that in order to undertake an action with net social benefits, those in the minority who are harmed must be paid. This amounts to redistributing some of the net benefits established by a policy to those individuals who are most hurt by that policy. Siting of noxious facilities tends to concentrate costs within an area proximate to the site while providing diffuse benefits over a wide area. Compensation measures can rearrange this distribution of costs and benefits. Lawrence Slaski has pointed out that: "Payoffs are important in that they help to spread the cost of development over a larger area than the area directly impacted. . . .[62]

From the losers' point of view, such payments are generally desirable insofar as they reduce or elminate the losses that would otherwise accrue from siting. The response of many who stand to lose from a collective action, according to Bruce Ackerman, is the view that:

" 'Yes, I know what you're doing is in the public interest, but why should I be singled out to bear the financial burden of a measure that will benefit the general public?' In other words, 'Pay me.' "[63] Ackerman alludes only to financial recompense, but this form of compensation is not likely to prove sufficient in hazardous waste facility siting. As Farkas has indicated: "For some, no price will be high enough to compensate for their perception of the potential dangers."[64]

In practice, it is very difficult to make a community no worse off than it was before the waste facility was located in its midst. Many communities distrust the compensation principle, remaining skeptical that they really will be compensated for the full range of negative impacts. Therefore, for compensation to prove successful as a strategic device in siting, traditional notions will have to be expanded. In-kind compensation techniques, for example, have the advantage of directly offsetting the costs associated with a new facility, and thus may be more important to those individuals in a community who are most affected by the facility and who might not be reached adequately by monetary payments to the local government as a whole. Again, however, this form of compensation best addresses tangible impacts; examples include restoration of property affected by a waste facility, repaving of roads, replacement of water supply wells, and provision of new fire-fighting equipment to a community's fire department.

New Jersey's siting bill established a 5 percent gross receipts tax on facilities to be paid to host municipalities to compensate for the additional, tangible expenses incurred by a town in which a hazardous waste facility is placed. The range of costs to be covered by this tax is extremely limited; clearly, there is no guarantee that residents will not be made worse off overall by the facility. In this respect, "strict liability" is perhaps the best form of basic compensation that exists in the New Jersey statute (although, because of the traditional legal conception of compensation as purely monetary, the bill considers only the gross receipts tax as "compensation"). Strict liability allows those who can show that they were harmed by a hazardous waste management facility (show of causality) to recover damages, without having to first establish negligence on the part of the facility operator. The strict liability provision was included after considerable lobbying by environmental and public interest groups. The justification of strict liability, according to the New Jersey Public Interest Research Group (PIRG), is that:

> Industry, by generating a hazardous waste facility, creates a risk to people living in the immediate area. These people do not benefit from that risk; industry does. . . . Strict liability says that if an industry is to make a profit based on the chance it is taking with other people's lives, then let industry pay if the risk that it has created does indeed cause harm.[65]

Measures like strict liability are designed to compensate for risk, and thus begin to cover some of the intangible costs of operating hazardous waste facilities.

A number of siting examples, both for waste disposal activities and for major energy facilities, suggest that citizens and local governments may be willing to some degree to balance their environmental and public health concerns against the perceived economic value of the proposed new facility: jobs, tax payments, a catalyst for other economic growth in the area, or even as a prerequisite to retaining industry already there.[66] Local governments in some instances may be induced to support a new waste facility because of its potential contribution to the community's fiscal well-being, although this may be the exception, not the rule. Local government officials in New Jersey's Jackson Township, for example, accepted land for a new municipal landfill in their community because of the tipping fees the town would receive. The site they selected was disastrous by today's criteria: highly permeable soil already damaged by mining; astride an important groundwater aquifer; near residences using well water. In 1979, pollution from this leaking site reached dangerous proportions. Wells were closed, drinking water had to be brought in by truck, and the state's Department of Environmental Protection was called in to clean up the mess (see Appendix A).

Despite such examples, some communities still seek the apparent economic benefits of hosting waste disposal operations. Newark's municipal government officials have expressed a willingness to accept a new waste disposal facility, primarily to obtain high tax payments. The city lobbied for inclusion of a 10 percent gross receipts tax in state siting legislation. As one of Mayor Kenneth Gibson's aides said in a 1980 State Senate Committee hearing, "Such economic incentives are important to give municipalities a true incentive to step forward to offer sites for the safe processing and destruction of hazardous wastes."[67]

Given the heightened public concerns about hazardous waste facilities so evident in the 1980s, however, more and more people seem to be expressing the feeling that the positive economic impacts of new facilities will pale in comparison to their potential dangers and resultant costs of many types. This is especially true for landfills, even those deemed "secure"; however, more sophisticated waste management facilities will likely face this attitude as well. Moreover, citizens are beginning to question the fiscal emphasis of their local officials, especially when public health is at stake.

Conceptually, at least, the degree of compensation appropriate to a particular siting proposal has two potential upper limits: one, that point at which compensation becomes so great that the facility is no longer economically acceptable to its developer; and two, the amount of money needed actually to purchase—at fair market value, or even above—all the homes and land in the area. That is, people who are unwilling to live near a particular facility might be offered sufficient compensation simply to move away. While they have obvious family and economic ties to the community, at *some* price people would be willing to move. In the final analysis, negotiations over compensation agreements for hazardous waste management facilities may cause people to ask themselves "at what price would I agree to move?" while facility developers ask themselves "do competitive conditions allow me to pay that much?" and "at this price, are other sites now more attractive?"

As with other aspects of hazardous waste facility siting, adequate economic compensation may be a necessary—but far from sufficient—condition for local acceptance. Such compensation is important even if people want the facility, to reallocate the overall costs and benefits. Where local support is lacking, however, in most cases such payments will not be sufficient to gain a local endorsement. Even with an attractive financial award, and a balanced siting process, many communities may still say "no." Some residents will be fearful of possible dangers to public health and of ultimate economic liabilities. Others may be angered that the offer of compensation resembles a none-too-subtle form of bribery designed to gain local siting approval, often made to poorer communities more susceptible to the lure. This bribery perception is apparent even when payments are designed just to make the community no worse off than before, with no conscious attempt to "buy" support.

It must be born in mind, however, that relatively few new sites will be needed nationally for waste facilities—not zero, but not a high number either. Therefore, even though many communities may still say 'no," even with good compensation schemes and a balanced siting process, this may be an acceptable public policy outcome. It is entirely appropriate that only a few communities conceived as possible locations would be willing to accept waste facilities. Indeed, one important index of having chosen the wrong community would be that no financial arrangement could be negotiated that would make the facility acceptable to local residents. In sum, no siting process should be judged on its ability to induce any group of people in any community to be willing to accept any facility, or to make it possible to build a facility anywhere some developer proposed to do so. One important function of a good siting process is to filter out those proposals that may look good at first but, upon further examination, are not attractive as sites. It is also essential to the political acceptability of any siting process that it be seen as able to defeat bad siting proposals.[68]

The standard practice in the United States of compensating only the one community that has land use jurisdiction over the site is too limited an approach, since neighboring communities may share the same water basin, airshed, underground aquifer, or transportation routes. Regional tax-sharing mechanisms may be needed to deal with this problem. Such mechanisms are employed for new land uses in Minnesota's Twin Cities regional area and in New Jersey's Hackensack Meadowlands, such that a number of communities share in the tax benefits of a new facility located in any one of them. The Massachusetts hazardous waste facility-siting statute allows nearby communities to be included in the Local Review Committee which plays a key role in the siting process. In the end, of course, it is difficult to balance equity (regional sharing of the revenues) against pragmatism (local focus, within the community whose land-use approval is desired).

The importance of regionwide acceptance of the proposed facility, as opposed to acceptance within the local community alone, remains unclear. Energy facility-siting controversies suggest that high tax payments may induce a local government and many of its constituents to accept the facility; in contrast, neighboring communities gain none of the tax revenues and may oppose the plant.[g] Available evidence does not, however, permit determination of whether public opposition to a new hazardous waste facility from the neighboring region

[g]This was the situation at New Jersey's Salem/Hope Creek complex of four nuclear reactors. All the gross receipts and franchise tax revenues go to the immediate host community, Lower Alloways Creek Township. Opposition from the surrounding communities has caused the state legislature to explore possible changes to gross receipts tax procedures. In Georgia, however, an opposite case may be found. While all of the tax payments from the Hatch nuclear generating station go to Appling County, the plant receives strong popular support from residents of adjacent Toombs County as well, due to the facility's overall value as a growth-stimulus to the whole area.[69]

would be sufficient to defeat a siting proposal. One needs to distinguish between legal ability to defeat it (nil) and political capability to do so (perhaps substantial).

For compensation to be an effective way to reduce the perceived costs of hazardous waste management facilities, it will be necessary to use a variety of mechanisms like strict liability that are aimed at reducing or eliminating the full range of negative impacts that accompany such facilities. In-kind replacement or restoration actions are compensatory strategies that can offset costs by replacing affected resources directly. For example, a developer might agree to provide new drinking water supplies if local groundwater becomes contaminated.[70]

Perhaps the most important form of compensation, however, is that designed to deal with risk: contingency funds and insurance. Contingency funds may provide some assurance that compensation will be available should an adverse event actually occur.[71] Such a fund might be established by a facility developer to repay property owners for their losses if property values declined as a consequence of a facility's impacts.[72]

Innovative use of basic insurance principles deserves a much greater role in siting hazardous waste management facilities, especially since insurance companies tend to scrutinize their clients' performance and alter their premiums accordingly. David Goetze, for one, has been exploring the possibility of offering one of two kinds of insurance to local residents living around a hazardous waste facility: insurance that the facility will indeed harm the environment—or that it will not. Premiums would be based in large measure on probability estimates made by the facility's developer.[73] However, such funds may do little to persuade people to accept willingly any risks that concern serious long-term public health.

The Issue of Incentives

Compensation, as noted, allows redistribution of the costs incurred by siting, using the net overall benefits from establishing a hazardous waste management facility to guarantee that no one is made worse off by these facilities. A community can thus share in the benefits of siting; rather than causing substantial losses to local interests, siting might then offer a "win-win situation."[74] In practice, mitigation and compensation techniques attempt to "recreate the status quo."[75] For individuals truly to be no worse off as a consequence of siting, however, measures would have to be taken which go beyond the facility's specific impacts. These actions have become known as "incentives": "Unlike compensation which reimburses communities for actual costs incurred, incentives are positive inducements to encourage communities to accept a hazardous waste facility."[76] In theory, incentives allow the developer of a facility to increase the benefits offered to a community and thereby enhance the attractiveness of his proposal, making it less of a bad bargain to local residents. This approach aims at use of positive rather than reactive measures which might frequently take the form of community improvements. They could include establishment of recreational facilities, perhaps, or funding a cleanup of existing abandoned hazardous waste dumps in the area.[77] At times, developers may use incentives to demonstrate their goodwill toward a community. Donation of equipment to a community is a common incentive. In siting and operating its facility in Idaho, Wes-Con, Inc., used such incentives as donations to local charities and provision of free pesticide disposal to area farmers.[78]

While provision of such incentives by a developer may facilitate local acceptance, such offers may backfire if they appear to constitute bribery.[79] This is particularly true when the incentives being offered are unrelated to a facility's specific impacts. In this sense, incentives suggested by the recipient community as part of a bargaining process may have a greater chance of success than do incentives suggested by the facility's sponsor.[80] William Ahern has also pointed out that offers of incentives may call added attention to the dangers associated with the particular facility. In the case of attempted siting of a Liquefied Natural Gas terminal in California, monetary payments might have proven a useful way to compensate people for additional risk. Such payments were not proposed, however, because the developer realized that they would "imply some serious or unacceptable risk" from the terminal.[81]

Unfortunately, the neat conceptual distinction between compensation and incentives breaks down in practice. Outside of the negotiated compensation agreement itself, there is no unambiguous mechanism by which to determine just what the total impacts of a waste facility are, and what they are worth to local residents. To them, the new park may simply compensate for the lessened community image. Thus incentives become subsumed in practice under the broader rubric of compensation.[82]

High financial compensation to the host community is not necessarily bad. If these payments win public approval while not sacrificing environmental and public health safeguards, they may indeed be desirable. This caveat is crucial, of course; one of the major problems with high compensation is the tendency it brings to divert attention to financial matters, and thereby induce people to overlook potentially adverse impacts on the environment.

How much compensation is enough, and who should set this figure? In Massachusetts, and as recommended in studies performed for the New England Regional Commission, compensation is determined flexibly through direct negotiations between the facility's sponsor and the host community. If compensation is set too low, local opposition may be overwhelming. This is well understood by all those concerned with siting hazardous waste facilities. The opposite situation is less clear in the literature, but may in the end be far more important. Setting compensation too high could have a negative impact on the profitability of operating the proposed waste facility; assuming that these costs are passed on to users of the facility, high local compensation by traditional methods could provide a further inducement for more illegal dumping of these wastes.

Illegal dumping is a particularly virulent and dangerous aspect of the hazardous waste problem. One disturbing fact is that stricter controls on legal disposal—at a high cost—may intensify inducement toward illegal disposal: "leaving the stopcock open at the back of the truck." Problems of effective enforcement are severe.[83] Perhaps fee schedules at new, environmentally sound facilities should be set low enough to encourage companies to use them. (States might even require that users receive a modest payment for each barrel or gallon or ton of waste that they bring to the facility, rather than having to pay to use it.) This "bottle bill" concept deserves careful study. John Mulvey has concluded that an "incentive-based system is a viable alternative to the usual direct regulatory approaches. It has been highly successful in several instances—bottles and waste oil, for example—and should be considered seriously by those interested in reducing improper disposal of hazardous substances, wherever they occur."[84] In debate on New Jersey's siting bill the proposed gross receipts tax on hazardous waste facilities was lowered from 10 percent to 5 percent per year due to concern over raising the price of legitimate waste disposal too high.

These concepts would require special financial subsidies of one kind or another, perhaps based on a tax on the industry at the point of waste generation (or on products with a high component of hazardous wastes) so that cost-free disposal would not constitute an incentive for increased production of these wastes. In theory, at least, it seems logical to tax waste generation instead of charging high rates for use of high-technology waste disposal facilities. Such taxes could produce the funds necessary to minimize illegal dumping (since legal facilities would be "free" or at low cost to the user), providing a marginal cost disincentive for waste production. However, what specific taxing procedures are to be preferred? What formulas make the most sense? Politically, these are not trivial questions. The theory calls for PCB's to be taxed at "x," for example, and fly ash at "y"; but specific values would have to be applied to *x* and *y*, and by whom? This would be complicated but might well merit further consideration.

Successful application of mitigation and compensation undoubtedly rests on using these techniques carefully in a sequential fashion during the siting process. The initial goal should generally be to mitigate the most undesirable and "costly" impacts associated with operation of a facility. As a Keystone Center report has pointed out, the first stage in the review of a facility proposal should be to assure that a sufficiently low "risk threshold" is obtained, thereby reducing public health risks to an acceptable level. Use of compensation for nonhealth-related impacts can accompany such mitigation.[85] After the whole array of impacts from a proposed hazardous waste facility have been dealt with in the siting decision, attention can be focused on the wider question of redistribution of benefits.

NEGOTIATING TO DETERMINE COMPENSATION

Mitigation, compensation, and incentives have significant strategic value in siting decisions inasmuch as these measures deal with the mismatch between costs and benefits which sustains opposition to siting. If the various community and regional interests who stand to lose from siting a facility are to be assured that they will not, in fact, be threatened, it is vital that these interests be allowed to participate in the decisionmaking process.

Negotiation, discussed previously as a preferred means of decision-making, is especially suited to the task of arriving at satisfactory agreements concerning appropriate compensation. Massachusetts' siting legislation, for example, specifies that a "siting agreement" be established between a facility developer and a host community. This amounts to a binding contract that entitles the host community to certain levels of cost reduction or benefit provision in return for its acceptance of the waste facility.[86]

These voluntary community-developer negotiations are supposed to produce agreement over the compensatory benefits important to the community's interests. State and federal regulations governing facility design and safe operation could establish a minimum level of actions (largely mitigation) which can serve as the basis for additional community-developer agreement. The very process of negotiating a satisfactory arrangement is likely to foster greater public acceptance, since those who are affected are able to assure themselves that the costs of hosting a facility would not be excessive.

Bargaining over compensation would make negotiations particularly desirable. The opposing interests of the developer and the community could be resolved, if possible, by agreement on a compromise set of provisions acceptable to all sides.[h] Negotiated decisions on hazardous waste siting, as argued in the previous chapter, should include a broad range of horizontal interests. If community officials bargain for an entire community, the result might be settlements which do not reflect the interests of many of the groups within the community. One of the biggest dangers is the tendency of public officials, observed in past siting attempts, to be concerned predominantly with the financial benefits that accrue to the town, whereas the residents of the town are likely to be most concerned about the risks of the facility and its safety features, not its property tax payments.

The biggest drawback to reliance on negotiated compensation is the possibility that local interests will refuse to negotiate. On some issues, in fact, community representatives may well reject the idea of compromise. In such cases, state-enforced arbitration and use of state override authority may be desirable to ensure that a balanced siting process can yield a final site determination, while also providing for a fair settlement. The National Governors Association, for example, has recommended that a state appeals board be used to accomplish this objective.[87]

[h] Regional interests must also be included in these negotiations as well, as indicated earlier.

Use of mitigation and compensation surely will not guarantee successful siting of hazardous waste management facilities. In many instances, however, greater attention to the potential of these measures is likely to spell the difference between acceptance and rejection of facilities. Consideration of how these techniques might best be incorporated in siting decisions has not been given adequate attention in most siting discussions. New Jersey's process, for instance, adopted a limited view of the importance of compensation.[88]

The fact that techniques like compensation have traditionally been tied to remunerative actions involving payments of money may help explain this reluctance. By dealing directly with the adverse impacts generated by facilities, however, nonmonetary means of compensation can provide an effective means to reduce opposition. By focusing compensation on specific public concerns, negotiation fulfills the goal of dealing with opposition to siting rather than bypassing it.

Notes

1. According to Katherine Montague, of the New Jersey Toxics Project, if people are aware of the need for hazardous waste management facilities, then they should be willing to accept such facilities being sited nearby. This, she admits, is an "optimistic" view of the siting problem. Discussion, Center for Energy and Environmental Studies, Princeton University, 21 November 1980.
2. One of the earliest papers to raise the compensation issue explicitly was Michael O'Hare, " 'Not On *My* Block You Don't': Facility Siting and the Strategic Importance of Compensation," *Public Policy* 25 (Fall 1977):407–458.
3. Booz, Allen & Hamilton, *Hazardous Waste Management Capacity Development in the State of New Jersey*, prepared for the State of New Jersey and the Delaware River Basin Commission (15 April 1980), p. IV–40.
4. Frank Michelman, "Property, Utility and Fairness: Comments on the Ethical Foundations of 'Just Compensation' Law," *Harvard Law Review* 80 (April 1967):1165–1167.
5. Fred Bosselman, David Callies, and John Banta, *The Taking Issue: A Study of the Constitutional Limits of Governmental Authority to Regulate the Use of Privately Owned Land Without Paying Compensation to the Owners* (Washington, D.C.: U.S. Government Printing Office, 1973).
6. David Ervin and James Fitch, "Evaluating Alternative Compensation and Recapture Techniques for Expanded Public Control of Land Use," *Natural Resources Journal* 19 (January 1979):22.

7. Michelman, "Property, Utility and Fairness," p. 1179. One could argue that "takings" law tends to compensate only those who are easily identified as being harmed by a public action and who are relatively few in number. More difficult means of compensation should not be avoided, however, simply because they are difficult to accomplish.
8. Ibid.
9. Malcolm Getz and Benjamin Walter, "Environmental Policy and Competitive Structure: Implications of the Hazardous Waste Management Program," *Policy Studies Journal* 9 (Winter 1980):410.
10. Gordon Bishop, "State Nearing Choice of Toxic Waste Sites," *Newark Star-Ledger* (8 December 1980), p. 1.
11. See State of California, Office of Appropriate Technology, *Alternatives to the Land Disposal of Hazardous Wastes: An Assessment for California* (Sacramento: OAT, 1981).
12. R. Kenneth Godwin and W. Bruce Shepard, "State Land Use Policies: Winners and Losers," *Environmental Law* 5 (Spring 1975):714.
13. Melvin Levin, Jerome Rose, and Joseph Slavet, *New Approaches to State Land-Use Policies* (Lexington, Mass.: Lexington Books, 1974), p. 68.
14. Diane Graves, Conservation Chairman, New Jersey Sierra Club, argues that those involved in organizational activities must have the resources to allow them to do so. Interview, 19 March 1981. In a self-perpetuating manner, then, economic advantages allow individuals to establish political strength.
15. National Governors Association, Energy and Natural Resources Program, *Siting Hazardous Waste Facilities*, final report of the National Governors Association Subcommittee on the Environment (Washington, D.C., March 1981), p. 1.
16. Ervin and Fitch, "Evaluating Alternative Compensation and Recapture Techniques for Expanded Public Control of Land Use," p. 36.
17. U.S. Environmental Protection Agency, *Siting of Hazardous Waste Management Facilities and Public Opposition*, SW-809 (Washington, D.C.: U.S. EPA, November 1979), p. 24.
18. Alan Farkas, "Overcoming Public Opposition to the Establishment of New Hazardous Waste Disposal Sites," *Capital University Law Review*, 9 (Spring 1980):452–454.
19. Michael O'Hare, personal communication to David Morell, 19 January 1982.
20. Robert Healy and John Rosenberg, *Land Use and the States,* 2nd ed. (Baltimore: Johns Hopkins University Press, 1979), p. 10.
21. Grace Singer, "People and Petrochemicals: Siting Controversies on the Urban Waterfront," in David Morell and Grace Singer, eds., *Refining the Waterfront: Alternative Energy Facility Policies for Urban*

Coastal Areas (Cambridge, Mass.: Oelgeschlager, Gunn & Hain, Publishers, 1980), pp. 19–63.

22. Sister Jacinta Fernandes, Testimony to the N.J. Senate Energy and Environment Committee in *Public Hearing on S-1300: Volume II* (Newark, New Jersey, 6 November 1980), p. 37.
23. Madelyn Hoffman, "Ironbound Resistance to Waste Incinerator," *Re:Sources*, Winter 1981–1982, p. 7.
24. Quoted by Singer, "People and Petrochemicals," p. 38; In cities like Elizabeth, residents are now beginning to ask the same questions about hazardous waste.
25. *Public Hearing on S-1300: Volume II* (Newark, New Jersey, 6 November 1980), p. 37.
26. National Governors Association, for example, recommends industrial sites. *Siting Hazardous Waste Facilities*, p. 5.
27. Ronald Luke, "Managing Community Acceptance of Major Industrial Projects," *Coastal Zone Management Journal* 7 (1980):279.
28. Healy and Rosenberg, *Land Use and the States*, p. 258.
29. U.S. EPA, *Siting of Hazardous Waste Management Facilities and Public Opposition*, p. 13.
30. John McCarroll, Testimony to N.J. Senate Energy and Environment Committee in *Public Hearing on S-1300: Volume I* (27 October 1980), p. 28A.
31. U.S. EPA, *Siting of Hazardous Waste Management Facilities and Public Opposition*, p. 49.
32. Ibid., p. 59.
33. Graves, Interview, 19 March 1981.
34. Michelman, "Property, Utility and Fairness," p. 1250.
35. U.S. EPA, *Siting of Hazardous Waste Management Facilities and Public Opposition*, pp. 54–59.
36. Ibid, pp. 38–44.
37. Richard Gimello, DEP Public Participation Chief, Conversation on S. 1300, 6 February 1981.
38. Booz, Allen & Hamilton, *Hazardous Waste Capacity Development in the State of New Jersey*, p. II–11.
39. "Minnesota proposes preferred areas for waste processing facilities," *Hazardous Materials Intelligence Report* (13 October 1981), p. 7.
40. Michael O'Hare, relating a conversation he had with Gordon Hester of the Minnesota Waste Management Board—personal communication, 19 January 1982.
41. See Ralph Keeney, *Siting Energy Facilities* (New York: Academic Press, 1980), pp. 80–112. An early approach to the use of multiple exclusionary factors in land-use planning is presented in Ian McHarg, *Design with Nature* (Garden City, N.Y.: Doubleday, 1969).

42. O'Hare, personal communication, 19 January 1982.
43. See Robert G. Healy, ed., *Protecting the Golden Shore: Lessons from the California Coastal Commissions* (Washington, D.C.: The Conservation Foundation, 1978).
44. Michael Baram, *Environmental Law and the Siting of Facilities: Issues in Land Use and Coastal Zone Management* (Cambridge, Mass.: Ballinger Publishing Co., 1976), p. 9.
45. New Jersey Hazardous Waste Advisory Commission, *Report of the Hazardous Waste Advisory Commission to Governor Brendan Byrne* (Trenton, N.J.: State of New Jersey, 1980), p. 38.
46. O'Hare, personal communication, 19 January 1982.
47. See EPA's *Siting of Hazardous Waste Management Facilities and Public Opposition*, p. 164.
48. Delaware River Basin Commission/New Jersey Department of Environmental Protection, Public Meeting on Level III Criteria for Identification and Screening of Sites for Hazardous Waste Facilities, Edison Township, N.J., 16 July 1980.
49. Interview with Diane Graves, Princeton, New Jersey, February 1982.
50. National Governors Association, *Siting Hazardous Waste Facilities*, p. 4.
51. Chapter 3 describes these costs.
52. Robert McMahon, Cindy Ernst, Ray Miyares, and Curtis Haymore, *Using Compensation and Incentives When Siting Hazardous Waste Management Facilities—A Handbook* (Washington, D.C.: U.S. EPA, 1982), p. 3. An earlier version of this report, in draft, was: Urban Systems Research and Engineering, Inc., *A Handbook for the States on the Use of Compensation and Incentives in the Siting of Hazardous Waste Management Facilities*, draft prepared for U.S. Environmental Protection Agency (Cambridge, Mass.: Urban Systems Research and Engineering, Inc., 30 September 1980).
53. United States, House, Committee on Interstate and Foreign Commerce, Subcommittee on Oversight and Investigation, *Hazardous Waste Disposal*, 96th Congress, First Session (Washington, D.C.: U.S. Government Printing Office, 1979), p. 19.
54. McMahon et al., *Using Compensation and Incentives*, p. 37.
55. Ibid.
56. Robert Bruce, *Mitigation, Compensation, and Incentives Used in the Siting of Hazardous Waste Facilities Under S-1300*, prepared for New Jersey Department of Environmental Protection (December 1980), p. 2.
57. Ibid.
58. McMahon et al., *Using Compensation and Incentives*, p. 7.
59. Tom Cook and James Knudson, *A History of Efforts to Acquire a Hazardous Waste Site in the State of Washington*, presented to the

Natural Resources Committee Meeting of the National Conference of State Legislatures (Denver, Colo., 15–16 February 1980), Appendix.

60. U.S. EPA, *Siting of Hazardous Waste Management Facilities and Public Opposition*, p. iv.
61. O'Hare, " 'Not on *My* Block You Don't'," p. 414.
62. Lawrence, Slaski, "Facility Siting and Locational Conflict Resolution," *Coastal Zone '78*, Volume 1 (New York: American Society of Civil Engineers, 1978), p. 16.
63. Bruce Ackerman, "The Jurisprudence of Just Compensation," *Environmental Law* 7 (Spring 1977):510.
64. Farkas, "Overcoming Public Opposition to the Establishment of New Hazardous Waste Disposal Sites," p. 459.
65. John Wilmer, N.J. PIRG, Testimony to Senate Energy and Environment Committee, Public hearing on S. 1300, Trenton, New Jersey, 17 December 1980.
66. The Monsanto/Bridgeport (New Jersey) case described in EPA, *Siting of . . .* (pp. 34–35), has certain aspects of this situation. Newark (New Jersey)'s desire for a new hazardous waste facility follows the same pattern.
67. Frank Sudol, Testimony at State Senate hearing on S. 1300, November 1980, p. 2.
68. Michael O'Hare was instrumental in making these distinctions explicit—personal communication, 19 January 1982.
69. Mark A. Shields, J. Tadlock Cowan, and David J. Bjornstad, *Socioeconomic Impacts of Nuclear Power Plants: A Paired Comparison of Operating Facilities*, NUREG/CR-0916 (Oak Ridge, T.N.: Oak Ridge National Laboratory, 1979).
70. James McCarthy, one of the victims of the Jackson Township well contamination, advocated this type of measure at the final public hearing on S. 1300 on 17 December 1980.
71. McMahon et al., *Using Compensation and Incentives*, p. 11.
72. A common fear in siting is of loss of property value. Though over the long run a well-run facility may not result in any losses, immediately upon being designated a site surrounding property values may drop and these residents' homes will become less marketable. Though this may be a temporary phenomenon, it is of great concern to many people faced with siting proposals. Graves, Interview, 19 March 1981.
73. David Goetze, "A Decentralized Mechanism for Siting Hazardous Waste Disposal Facilities" (Washington, D.C.: Resources for the Future, n.d.). (Mimeo) The General Research Corporation of McLean, Virginia, has been conducting risk-assessment studies for a small but growing number of insurance companies which offer insurance to owners and operators of hazardous waste facilities. Presentation by Douglas Britt, General Research Corporation to National Con-

ference on "Groundwater in the '80's" (Chicago, 12 November 1981).

74. Clark-McGlennon Associates, *A Decision Guide for Siting Acceptable Hazardous Waste Facilities in New England*, prepared for the New England Regional Commission (November 1980), p. 8.
75. Lawrence Bacow, *Mitigation, Compensation, Incentives and Preemption*, prepared for the National Governors Association (10 November 1980), p. 5.
76. Ibid., p. 8.
77. Urban Systems Research & Engineering, *A Handbook for the States*, pp. 73–74.
78. U.S. EPA, *Siting of Hazardous Waste Management Facilities and Public Opposition*, pp. 144–156.
79. Urban Systems Research & Engineering, *A Handbook for the States*, p. 8.
80. Ibid., and in other parts as well.
81. William Ahern, "California Meets the LNG Terminal," *Coastal Zone Management Journal* 7 (1980):218. Also see Lawrence Susskind and Stephen Cassella, "The Dangers of Preemptive Legislation: The Case of LNG Facility Siting in California," *Environmental Impact Assessment Review* 1 (1980):9–26.
82. O'Hare, personal communication, 19 January 1982.
83. See Michael O'Hare, "Enforcement and Source Reduction: Must We Subsidize Hazardous Waste?" Paper presented at National Conference on Hazardous Waste (Newark, N.J.: June 1980).
84. See John Mulvey, "An Incentive-Based Resource Recovery System: Reducing Improper Disposal of Hazardous Wastes" (Princeton, N.J.: School of Engineering and Applied Science, Princeton University, March 1981), p. 36.
85. Keystone Center, *Siting Non-radioactive Hazardous Waste Management Facilities: An Overview*, final report of the First Keystone Workshop on Managing Non-Radioactive Hazardous Wastes (Keystone, Colo.: The Keystone Center, September 1980), p. 23.
86. Commonwealth of Massachusetts, *Massachusetts Hazardous Waste Facility Siting Act*, sec. 12.
87. National Governors Association, *Siting Hazardous Waste Facilities*, p. 10.
88. As used in the New Jersey process, the notion of compensation has been limited to financial measures such as the 5 percent gross receipts tax.

6 CONCLUSIONS: PUBLIC POLICY AND THE MYTH OF PREEMPTION

It must be recognized that the public may never accept hazardous waste disposal sites. There may be absolutely no way to convince the average citizen that a well-designed, -operated, and -regulated site poses minimal risk, especially compared to illegal dumping.

Robert Pojasek (1980)[1]

Negotiations over facility design, mitigation and compensation can help take into account those costs that have previously been overlooked. In so doing, communities can begin to turn these facilities into an asset to be sought in place of an undesirable facility to be opposed.

Clark-McGlennon Associates (1980)[2]

The notion that communities might come to view hazardous waste facilities as an "asset" seems rather absurd when contrasted with prevailing public sentiments toward these projects. Although public opposition to facilities may well be diminished somewhat by reliance on negotiated compensation, it is evident that not everyone's views can be "realigned" with the broader public interest. Even the very best siting policy will probably continue to draw protests from some, particularly from residents of the immediate area surrounding the proposed new facility.[3]

Though the view of waste facilities as assets seems overly sanguine, a pessimistic view like Robert Pojasek's does not seem fully warranted either. Instead, the premise behind siting policy should be that the difficult political issues that attend this problem can be dealt with in a fair and effective manner, thereby leading to broad citizen respect for the siting process and thus to successful establishment of needed new waste facilities. For this success to occur, however, policymakers must be willing to depart from the conventional assumptions and tactics which have contributed to the current siting impasse.

A "PARALYZED" STATUS QUO

The current situation with respect to siting hazardous waste facilities is unacceptable. Within the next decade, increased demand for hazardous waste disposal, coupled with insufficient disposal capacity, threatens to increase illegal disposal, a result which would exact a terrible price in terms of environmental degradation and danger to human health.[4] The past history of hazardous waste management has been nothing short of tragic; we can ill afford to carry on these practices.

The impediments to establishing the facilities needed for "safe" disposal are, however, considerable. Much of the intense public opposition that has thwarted siting can be attributed to an uneven distribution of costs and benefits: large per capita losses are concentrated on a particular locality or region in order to provide diffuse benefits for the entire state. Siting of hazardous waste management facilities, like other facilities with significant regional impacts, thus manifests an important redistributive aspect.

Not surprisingly, those who are the victims of these large localized losses have objected strenuously to facility-siting proposals. They have felt exploited and mistreated by the broader society and fearful for their families' health. And to the degree that local interests can control land-use decisions, these objections have been reflected in local vetoes of proposed facilities. Although home-rule procedures are desirable to allow citizens the largest possible measure of control over their own destiny, it is regrettable that these same procedures militate against attaining an urgent statewide and even national benefit: safer disposal of hazardous waste.

THE MYTH OF PREEMPTION AS AN "EASY" ROUTE

One means to try to overcome local opposition to siting of facilities is to legislate around it. This approach simply does not work, however.

Despite the political costs involved in making communities relinquish their existing power over land-use control, several states have responded to the siting problem by a show of political will on the part of state government via passage of preemptive siting policy. The logic of this tactic is that in the face of local resistance to actions of statewide benefit, the state should step in to ensure that sites can be obtained and facilities constructed. New Jersey's initial siting legislation (S.1300), for example, reflected this preemptive philosophy in establishing a state siting commission that could condemn any land in the state for hazardous waste facility sites without reference to any local zoning regulations or other local ordinances.

Though this strategy appears to connect the statewide need for new hazardous waste management facilities with an efficient mechanism for removing locally imposed obstacles, such preemptive power is *not* likely to prove workable.[5] Indeed, successful reallocations of authority should not be expected, given the American tradition of constraints on unpopular or unchecked exercise of power. Local interests will surely employ a broad range of tactics calculated to block or delay siting proposals. Even if preemptive authority does "work" in certain instances, these facilities will be sited only at a substantial cost in terms of community disaffection and individual alienation.[6]

Repeated failures to obtain sites for hazardous waste facilities have led some to conclude that the best siting strategy is to find sites where public opposition is lowest, generally in industrial areas or in very remote rural locales. This approach, designated the path of least resistance, can likewise be rejected on both strategic and equity grounds. The success of this plan is jeopardized by the fact that, increasingly, urban residents are asserting their own right to a safe environment and are expected to oppose future urban siting proposals.[7] This is surely the case in Elizabeth, New Jersey, for example, where the community group CUE has begun to oppose the continued presence of hazardous substances, undoubtedly in response to the awful consequences of illegal waste management practices in that city by the Chemical Control Corporation. Some resistance can also be anticipated in rural areas, for here land-use compatibility is lacking and the danger of transporting the wastes over long distances imposes another kind of high social cost.

A policy that requires certain segments of society to bear the largest share of the costs of our industrial economy is unfair as well. The presence of other industrial installations in urban areas should not condemn these residents endlessly to accommodate similar additional land uses. These people are correct when they complain of being "dumped on" when such policies are perpetuated.[8]

THE "DIFFICULT" ROUTE: BALANCE AND NEGOTIATION

In contrast to the apparently "easy" measures, mechanisms for balanced decisionmaking appear at first glance to be cumbersome, inefficient ways to obtain needed sites. Nevertheless, such procedures as a sequential override are far more likely to succeed in achieving the objective of establishing sites, provided such processes incorporate a fair balance between both state and local interests. While it may make sense not to give communities final decisionmaking authority (a complete "local veto") over siting, these local interests must be allowed to participate actively and authoritatively throughout the siting process. A full local "yes or no" decision on the proposed facility, followed by the possibility of a subsequent state override, is the best way to encourage such local citizen participation. In its absence, the legitimacy of the siting process will be in doubt, as will its ultimate success. Provisions can be included in state siting legislation to ensure that local decisions are made on a timely basis, and that state exercise of override authority must meet strict formal controls.

Balanced decisions should also respect differences in the horizontal dimension of siting. A potential abutter to a proposed facility will have a different viewpoint than will a potential employee of this facility. Both should be allowed to participate in the siting decision, since this decision has a profound importance for both of them.

Even a relatively "open" decisionmaking process will have the tendency to rely on organized (structurally strong) groups to represent differing interests. This bias suggests that siting policies, particularly when they involve low-income areas, should establish innovative techniques to respond to this problem. One possibility is that residents at least be surveyed as to their attitudes, particularly in the absence of any community groups that exist to speak for them in the siting negotiations. Such surveys would not ask whether the residents were simply for or against a proposed facility, but what their concerns were about its adverse impacts and what measures would be required to compensate them for these impacts.

The best means of arriving at an equitable hazardous waste siting decision appears to be through negotiation. Again, use of this strategy might be seen as cumbersome and time-consuming, but it should ultimately have the most beneficial results. Its use is motivated by the belief that those individuals and groups who oppose facilities should be made an integral part of the decisionmaking process. The goal is to address public concerns, not bypass them. Although advice-giving and error-detection modes of participation have some value as well, they are far more limited than negotiation in the amount of authority granted to different participants to pursue their own interests.

Realistically, some amount of state coercion will have to be present in siting to guard against local refusal to undertake these decisions which are needed by the broader community. The best means to accomplish this aim appears to be to provide for state-arbitrated settlements; interests will then have an incentive to participate in the negotiations for fear that otherwise the final settlement will not be favorable for them. Unlike preemptive strategies, this approach combines state coercion through override with deference to the local right to participate.

Despite these more legitimate means of selecting sites, it is still quite likely that siting will not overcome all local opposition. Unless siting policies reduce the extremely high level of costs that are bestowed by the act of siting, it is unfair and unrealistic to expect those who are harmed to consent to these actions.

From a strategic point of view as well, it is vital to the success of hazardous waste facility siting that techniques of mitigation and compensation be employed to reduce the net costs to those who must live near the new facility. This is accomplished by spreading out such costs more broadly among the beneficiaries of this siting action.[9] Although siting might ideally present a case of "win-win" policy by use of redistributive techniques, realistic constraints on these mechanisms will probably mean that siting still establishes winners and losers. As Robert Healy and John Rosenberg found:

> Unfortunately, even the most astute and sensitive planning will not be able to make everyone at least as well off as before. Some will gain and others lose, whether by unchecked development or by carefully conceived land policies.[10]

In sum, any approach to the siting dilemma which hopes to succeed must recognize that local communities have a de facto ability to obstruct unwanted facilities, and thereby to defeat most siting proposals by making them too costly financially for the developer to pursue and too costly politically for the state to insist upon their construction. Despite their extensive constitutional authority over land use, states simply do not have the power to legislate successfully against committed local opposition.

Veto power over siting is not held by the locals alone, however—mutual balance exists in the authority also available to facility developers and to the states. The developer can simply abandon his siting proposal if, for example, local demands for compensation are too high or state regulatory requirements are too onerous. And states have the legal authority to veto any siting proposal by refusing to grant the facility the necessary permits and licenses. Thus we see a sharing of the authority to prevent the project from proceeding to completion.

None of these parties, however, has the authority to cause a facility to be built in the face of determined opposition from any of the other parties.[11] Since a certain number of new waste treatment, storage, and disposal facilities will be needed in the years to come, those that will be built are the ones that can receive some sort of acceptance—however grudging—from the various parties at interest.[12] The task required of the siting process is to accomplish this objective.

This process must be perceived as legitimate by all the participants; it must be worthy of their respect. Such a goal requires a balanced, sequential, and timely mechanism for decisionmaking. In addition, facility siting proposals themselves, at least some of them, must be acceptable to the host community and its residents. This can best be accomplished by extensive mitigation of their adverse impacts accompanied by negotiated compensation. State access to carefully constrained override powers can limit disgruntled community obstruction and ensure open bargaining, while explicit recognition of de facto local veto power can ensure that the host community can bargain from a position of appropriate power. A siting policy that aims to reduce inevitable losses to an acceptable level for those who are harmed and to build citizen respect through balanced, sequential, and timely procedures is not only carefully conceived but, more importantly, should also prove effective in clearing the political impediments to successful siting in the United States of an adequate number of the needed new hazardous waste management facilities. This structure is in accord with the political realities of hazardous waste siting controversies, and very different from the myth of preemption.

NOTES

1. Robert Pojasek, "Developing Solutions to Hazardous Waste Problems," *Environmental Science and Technology* 14 (August 1980): 929.
2. Clark-McGlennon Associates, *Negotiating to Protect Your Interests: A Handbook on Siting Acceptable Hazardous Waste Facilities in New England,* prepared for New England Regional Commission (Boston, Mass.: November 1980), p. 45.
3. According to Diane Graves, Conservation Chairman, New Jersey Sierra Club, the realistic goal of siting policy is not to gain everyone's approval but simply to "lower the decibel level" of the protests. Graves, Interview, 19 March 1981.
4. This research has not investigated the disturbing possibility that by establishing increased waste disposal capacity, illegal dumping will continue, or even *increase,* due to higher disposal costs. See, for example, Malcolm Getz and Benjamin Walter, "Environmental Policy and Competitive Structure: Implications of the Hazardous Waste Management Program," *Policy Studies Journal* 9 (Winter 1980): 408-412. Also see a survey of waste haulers, conducted for East Windsor Township, New Jersey—RL Associates, *Illegal Disposal of Hazardous Wastes: A Survey of Licensed Haulers* (Princeton, N.J.: RL Associates, 1980). If these incentives for illegal disposal of wastes were changed, new waste disposal facilities would be required.
5. New York State, for example, has encountered severe resistance to its new preemptive siting law. Residents of the Sterling area in Cayuga County, near Lake Ontario, have protested the site proposed for their area. Lena Williams, "New York State Finds a Site for Toxic Waste Plant," *New York Times* (13 May 1981): B-4, and "Town Fights Plan for Toxic Waste Treatment Plant," *New York Times* (31 May 1981): 35.
6. In New Jersey, this situation is most likely to occur along northern-southern lines, as just mentioned. A movement is already underway to have the southern half of the state become a separate state. If southern New Jersey communities were to become involved in a preemptive siting process which intended to site waste generated in the northern part, this separatist sentiment could conceivably grow stronger.
7. See Grace Singer, "People and Petrochemicals: Siting Controversies on the Urban Waterfront," in David Morell and Grace Singer, eds.,

Refining the Waterfront: Alternative Energy Facility Policies for Urban Coastal Areas (Cambridge, Mass.: Oelgeschlager, Gunn & Hain, Publishers, 1980), pp. 19–63.
8. See Morell and Singer, *Refining the Waterfront,* in various parts.
9. Ideally, the costs of hazardous waste disposal would be passed along to the consumers of hazardous waste products.
10. Robert Healy and John Rosenberg, *Land Use and the States,* 2nd ed. (Baltimore: Johns Hopkins University Press, 1979), p. 242.
11. As Judge Hall wrote in the precedent-setting New Jersey case on exclusionary zoning: "courts don't build houses." *Southern Burlington County N.A.A.C.P. v. Township of Mount Laurel,* 67 N.J. 151, 336 A. 2d 713 (1975).
12. Comments from Michael O'Hare were helpful in pointing to this issue—personal communication to David Morell, 19 January 1982.

APPENDICES

APPENDIX A

A CASE STUDY OF VICTIM-CENTERED POLITICAL ACTION: JACKSON TOWNSHIP, NEW JERSEY

David Moldenhauer

The politics associated with victim-oriented response to leaks from toxic waste landfills provide critical insights and help define the debate over siting new hazardous waste management facilities. One key difference should be noted at the outset, however: unlike the opponents in the various siting controversies, toxic waste victims make up only a small proportion of their community. Most residents of Niagara Falls, New York (Love Canal), of Elizabeth, New Jersey (Chemical Control Corporation), and of Jackson Township, New Jersey, for example, have not been affected directly by these wastes. The different responses to the tragedy within the community clearly reflect this fact.

Jackson Township lies in Ocean County, in the southeastern part of coastal New Jersey. Once a decidedly rural area, Jackson has quickly become suburbanized. In 1950, its population stood at 3,515. At that time, the township's economy was almost totally agrarian, depending to a large extent on chicken farming. As the economy of coastal New Jersey grew, so did Jackson. Many of the new residents worked in other communities of Ocean County, or in Monmouth, Mercer, or Union Counties. In 1960 the township's population was 5,939; in 1970, 18,276; and in 1980, about 25,000. All told, Jackson's population grew at an average rate of 7.2 percent per year between 1960 and 1980. This growth in population was accompanied by a flurry of new housing. In 1970, for example, there were about 5,000 housing units; in 1980, there were about 8,000.[1]

The township provides neither central sewage service nor drinking water throughout most of its 100-square-mile area. Instead, local residents rely on septic tanks and private drinking water wells. Nonetheless, Jackson's high rate of population and housing growth, unaccompanied by corresponding growth in business and industry, has created something of a fiscal crisis. A plan written in 1964 noted, for example:

> The population boom commencing in south-central New Jersey could well be disasterous to any municipality without a well-balanced economic and tax base. . . . a progressive community needs a balanced economy. Jackson's urban core and its large future population will undoubtedly support a regional shopping area and a healthy overall commercial economy within the Township's own boundaries. But otherwise the entire community will be residential unless the industrial areas are developed.[2]

Partially to alleviate this problem, the plan suggested that the township's southeastern corner be used for additional industry. Jackson's attempts to encourage development in this area, unfortunately, were to have severe environmental consequences.

In 1970, Jackson officials closed the township's municipal landfill in the Robins Estates area because of complaints from nearby residents that this landfill was threatening their health. As later reported in the *Asbury Park Press:*

> Residents of the Robins Estates development complained to township officials that the landfill operation on Hulse's Road polluted their well water, attracted rodents and seagulls and created dust and odor. Although it was never proved that the landfill operation created the water problems, the township closed the landfill and extended municipal water lines to the development.[3]

For more than a year, the township was without a municipal landfill. Jackson sent its wastes to a dump in another township. In 1971, however, the Glidden-Durkee Division of SCM Corporation offered Jackson 125 acres to use as a new landfill. This site was in the southeastern part of the township, near the Legler subdivision (see Figure A-1). Glidden had mined this area for clay, minerals, and dirt to use as fill at its various construction sites. The company decided to give the land to the township in order to avoid paying some $2 million for reclamation as was required by township ordinance. Jackson waived

Figure A-1. Location of the Jackson Landfill.

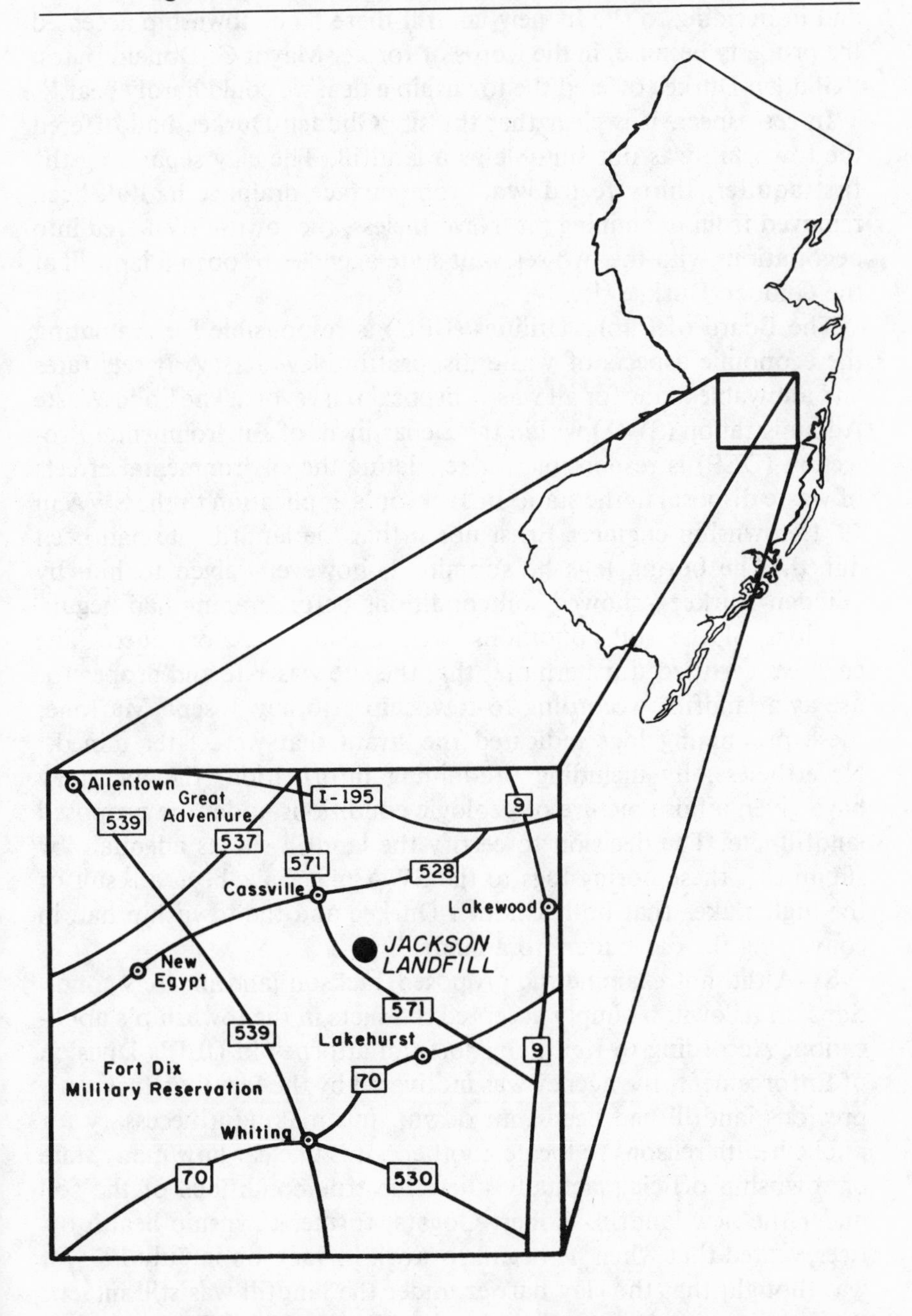

this provision in its municipal code in accepting the property donation and in deciding to site its new landfill there.[4] The township accepted the property because, in the words of former Mayor G. Donald Bates: "Glidden-Durkee offered the township a deal we could hardly beat."[5]

In retrospect, it is clear that the site Glidden-Durkee had offered the township was not suitable as a landfill. The clay separating the first aquifer, thirty feet down, from surface drainage had all been removed from the mining pit. Nevertheless, the township entered into negotiations with the two relevant state agencies to open a landfill at the Glidden-Durkee site.

The Board of Public Utilities (BPU) is responsible for regulating the economic aspects of waste disposal in New Jersey. It sets rates and allowable loads for all waste disposal purveyors. The Solid Waste Administration (SWA) within the Department of Environmental Protection (DEP) is responsible for regulating the environmental effects of waste disposal in the state. In Jackson's application to the SWA in 1971, township engineer Ernst noted that the landfill site had been mined. The boring logs he submitted, however—given to him by Glidden-Durkee—showed soil conditions before mining had begun. No logs of the soil conditions after mining were provided. The engineer certified, furthermore, that the site was safe and proper for use as a landfill. According to township attorney Joseph Martone, these pre-mining logs indicated the strata that were later mined.[6] Nevertheless, by including pre-mining information, the logs may have given a false picture of geologic conditions under the proposed landfill site. The decision to certify the landfill site as adequate by submitting these boring logs to the SWA may have been a result of the high stakes that both Glidden-Durkee and the township had in converting the old mine into a landfill.

SWA did not examine the proposed Jackson landfill site's conditions on its own; it simply accepted the facts in the township's application. According to Keith Onsdorf, an attorney in DEP's Division of Enforcement, the agency was motivated by the fact that Jackson's previous landfill had been shut down, thus making it necessary for public health reasons to locate another.[7] It is unclear how many state or township officials actually knew the true conditions of the soil under the new landfill. Robert Gogats, former township health officer, stated that when he began to work in Jackson in July 1975, it was thought that the clay barrier under the landfill was still intact.[8] When contamination was first reported later in 1975, DEP announced publicly that the Legler site was one of the best in the state.

Residents of the Legler area objected to the siting of the new landfill in 1971 for exactly the same reasons that the Robins Estates residents had demanded that the old landfill be shut: namely, that the landfill would be a bacteriological hazard and a public nuisance. Residents formed a Concerned Citizens' Committee and gathered 319 signatures on a petition to the township (opposing approval of the new landfill). All their attempts to argue against the site were ignored by the township officials, however. Mayor Bates later described the reasons the Township Committee decided against the Legler residents: "I remember the people in the area objecting. They didn't want it in their backyard. . . . But no matter where we put the landfill, people would holler."[9] In addition, in 1971 the area around the landfill was not as densely populated as it would be several years later. According to James McCarthy, spokesman for the present victims' organization, the Legler residents recognized that they had shallow wells and pointed out to township officials that groundwater flowed under the proposed landfill site to their wells. These facts, however, did not induce township officials to investigate the Legler area.[10] Nor did officials control housing construction around the landfill after it was approved.

Siting of the landfill and later decisions on use of surrounding land therefore took place in an atmosphere of institutional diffusion of responsibility. The state left the certification process to the township engineer, while the township left consideration of possible environmental effects to the state, even though the state may not even have known of the area's true geological conditions. In the final analysis, the landfill was sited in the Legler area because the Glidden-Durkee site was the only one which township and state officials seriously considered.

Between 1972 and 1975, residents of this area remained unaware of the possible effects the landfill was having on their health through its contamination of drinking water. The state and township, on the other hand, encountered continual operating problems with the facility. While the BPU allowed up to 300,000 gallons per day of septic tank sludge to be dumped there, adequate amounts of solid refuse to absorb this amount of liquid wastes were not being delivered to the landfill during this period. As a result, the septic tank sludges turned into a pond, creating noxious odors and potential health hazards. These conditions violated state health regulations, and DEP and the state Health Department cited the township nineteen times between 1973 and 1978.[11] In 1976, after several previous attempts, the township successfully petitioned the BPU to reduce the amount of septic wastes allowed on the site to 20,000 gallons per day.

Economic factors played a major role in keeping the landfill open despite these problems. Over the eight years the landfill was in operation, the township collected $960,000 in fees from the disposal of septic wastes. The Jackson facility charged more than other landfills in the state for disposal of solid wastes ($20 per forty-cubic-year load) and less than other landfills for disposal of liquid wastes ($6 per 1,000 gallons). "Thus the haulers disposed of their liquid wastes at the Jackson landfill while their solid wastes were disposed of elsewhere."[12] Given relative prices, it should be no surprise that liquid septic wastes accumulated at the Jackson landfill.

At the same time, at least some of the transporters of these septic wastes apparently were illegally including toxic chemical wastes in their loads. According to the later pleading files of the Legler residents' lawsuit, "a major chemical fire occurred at the landfill in 1976, which was not responsive to conventional firefighting techniques, indicating the presence of chemical contaminants."[13] In addition to the illegal dumping of toxic wastes in the landfill, there were numerous waste spills in the area. Gogats reports that between 1975 and 1978 the township's Board of Health was informed of several chemical spills near the site, including one of "giant" proportions. Gogats informed the DEP of illegal chemical spills in the area and state and local officials attempted to identify illegal dumpers, but to no avail.[14] The township purportedly also knew that illegal dumping was going on at the landfill, and "had known since 1975 from required testings that dangerous chemicals were leaking from the landfill into the Cohansey aquifer."[15] Nevertheless, the township made no efforts to close the landfill, to inform nearby residents of the potential dangers, nor to restrict new housing construction in the Legler area.

In 1977, residents moving into new houses in the Legler area began complaining about foul water and health problems. Gogats reports that in February or March of 1977, two households living closest to the base of operations of a company transporting septic tank wastes began complaining of foul well water,[16] which had "an unpleasant and a sewage-like odor. They blamed the water for nausea and intestinal disorders."[17] The township's Board of Health investigated Legler area wells and found hydrogen sulfide in the well water, as well as changes in its pH. The township requested DEP to begin sampling well water, but the state agency replied that it had neither the funds nor the manpower to do so. The township also contacted its state legislator. But it was only in November of 1978, as reports of foul water reached the press, that the state began testing water supplies. On November 11, 1978, in response to further requests by the Legler residents, the Township Committee barred further dumping of sludge at the landfill.

On November 22, 1978, however, the BPU ordered the township to reopen its landfill to sludge dumping. BPU argued that the facility was registered with the state as a public utility, and thus the township lacked permission to close the dump on its own. Though lacking the authority to assess the environmental risks of contamination from the landfill, the BPU could compel the township to keep it open. Only DEP could close the dump for health reasons but at that time DEP's tests had not yet shown a hazardous level of biological or chemical contaminants in the groundwater.

The township began collecting evidence "in an attempt to convince the BPU that the landfill should be closed (to sludge)."[18] In December 1978, DEP tests began to show that various organic chemicals found in the water corresponded to chemicals found at the landfill. Township officials chided DEP in press reports for not having analyzed the well water quickly. According to township health officer Gogats, after the samples were taken, DEP refused to say what was in the water or whether the water was safe, saying that standards were not set for such chemicals. Even after organic chemicals were reported in the water, the BPU still refused to allow the township to close the landfill to sludge dumping. To gain BPU approval, the township would have to prove that these septic tank wastes were the source of the toxic chemicals found in the drinking water.[19]

On December 21, 1978, DEP declared a public health emergency in Jackson and closed the landfill to further sludge dumping. The BPU acquiesced to DEP's declaration. At this point the township began at its own expense to deliver replacement water in tanks to the 165 affected Legler area households.

By July 1979, DEP had defined the extent of the contaminated area. The agency restricted its subsequent water tests to that area. The state claims that the extent of the groundwater contamination has not expanded since that date. A 1980 hydrological study noted, however, that the aquifer moves quickly, carrying slugs of organic chemicals with it.[20] McCarthy also claims that residents outside of the contaminated area show physical symptoms similar to those experienced by the Legler residents.[21] According to the *Asbury Park Press,* the township health officer attempted without success to persuade DEP to broaden its investigation. Gogats feared that the size of the contaminated area might be larger, but there was no way for the township itself to study this issue. The township could only plan a replacement water system for the 165 households in the four-square-mile area determined by the state to be contaminated.[22] Planning for replacement water was also contingent on approval from the state Pinelands Commission of the new deep wells being drilled as a replacement water source for the Legler area.

The total cost of the replacement water system has been estimated at $1.2 million. This includes the deep wells, pumps, water mains, and individual hookups. Normally provision of drinking water is the responsibility of the locality. In January 1979, however, Jackson Township requested disaster funds from the state to construct these replacement water supplies. Representatives from Ocean County introduced a bill in the state legislature authorizing a $1.2 million (5 percent) state loan to Jackson to help construct deep well replacement water supplies. In October 1979, the Township Committee passed a $650,000 bond ordinance to pay for construction of part of the water system and granted a construction contract for $503,759; Governor Brendan Byrne then approved the state loan to Jackson. The Township Committee also passed a mandatory hookup ordinance which required all of the 165 residents to connect to the planned water system at rates set by the township. Although Jackson has used a mandatory hookup ordinance in the past,[23] in this case, Jackson passed the ordinance before responsibility for replacing water supplies had been determined in the courts. Furthermore, thirty-five of the Legler families had access to private deep wells which supplied them with clean water. Martone states that the township decided to require all Legler residents to hook into the municipal system in order to reduce the cost of the system to township residents.[24]

In the fall of 1979, DEP ordered the township to begin construction of a replacement water system, with completion slated for February 1, 1980. In January 1980, with construction of the system not yet begun, the judge in a DEP suit to close and seal the Jackson landfill ordered the township to finish the system by June 1. On July 30, 1980, the new system was in place and residents began hooking into it. According to Jim McCarthy, the township originally intended that the Legler residents reimburse the municipality for the entire costs of the replacement water system.[25] The *Asbury Park Press* reported: "Township Attorney Joseph F. Martone and Mayor Donald Bates say the township has no legal responsibility to provide residents with deeper wells. 'But, morally, I think the township would be responsible,' Bates said."[26] By the time the replacement system was in place, the township had realized that its users could not bear the entire cost of the system, which was approximately $7,270 per house. The township therefore set a $500 connection fee for each of the 165 homes, and set water rates at twice those typical in this area.

By mid-September 1980, 103 of the 165 Legler households had connected to the new water system on these terms. Two-thirds of the others drew their water from new deep community wells put in by developers or by the residents themselves. The water from these wells was tested and found pure. Five or six other families were "die-hards" who, out of principle, would not hook into the municipal system. Another fifteen or so families could not hook into the municipal system because of financial hardship.[27] According to the hookup law, however, all of these sixty-two households were required to attach to the municipal system. According to Township Attorney Martone, the township will lose $60,000 per year in operating the system even when all of the residents are hooked in; with fewer residents using the system the township would lose even more.[28] On September 15, 1980, the Township Committee passed a resolution giving the remaining residents thirty days to connect into the system or face fines or imprisonment. On November 19, the township announced it was halting deliveries of replacement water to the sixty-two remaining homes. Martone later stated that this decision to cut off water supplies "was based on a misunderstanding." In his view not all of the local officials had wanted to cut off the water deliveries. In order to keep deliveries, however, the Legler residents had to obtain a temporary court injunction incumbent on their showing financial hardship.[29]

Jackson Township has never formally acknowledged that its landfill is the source of the chemical contamination in the aquifer. Indeed, DEP has identified other illegal toxic waste dumps in the vicinity of the landfill as secondary sources of groundwater contamination. The township suggests other possible sources:

> It is commonly known and recognized that such shallow wells obtain their water supplies from surface water runoff and surface water percolation, and that such surface waters are subject to contamination from

> many sources, including septic tank overflow, septic system leaching beds, automotive gasoline and oil and emissions, and illegal dumping of numerous types of waste products.[30]

Even if toxic wastes had been illegally dumped in the municipal landfill along with septic tank wastes, the township denies negligence. Martone asked: "Were we supposed to not allow the tank trucks to dump until we had taken a sample of each one, sent it off to be analyzed at $300 apiece, and waited two weeks for the results?"[31] Nonetheless, because DEP's landfill permit prohibited disposal of hazardous wastes, the township was responsible for ensuring that hazardous wastes were not dumped there. The township also denies negligence for allowing the Legler residents to drink the polluted water.

> As of the date that Certificates of Occupancy were issued to plaintiffs. . . defendant had no choice but to issue such Certificates of Occupancy for the reason that the dwellings and the potable water supply were in full compliance with all state and local requirements applicable thereto.[32]

Obviously, the township is sensitive to accusations that it is liable for the health and property damage incurred by the residents.

In October 1979, after ninety-six homeowners filed their intention to sue the township for health effects and property damage, the township filed a countersuit. The *Asbury Park Press* described the town's counterclaims:

> One claim would have the court revoke the certificates of occupancy for the home if the court rules the township granted the certificates improperly because the water was polluted. The other claim asks that residents of the Legler section who took part in the suit pay back the township $3,000 a month for salaries and wages of township employees and $10,000 for a truck and equipment used in delivering water to 160 (*sic*) homes since the water supply was found to be polluted a year ago.[33]

Martone stated that the purpose of the counterclaim was to force victims who hadn't hooked into the municipal water system to do so by revoking their certificates of occupancy, and to gain a "setoff" from any decision the court might make for the Legler victims. The counterclaim is to be heard at the same time as the victims' suit.[34] The claim itself, however, requests compensation from the victims, no matter how their own claim is decided. Arnold Lakind, the victims' attorney, argues that the counterclaim is one way the township is showing vindictiveness toward the Legler residents.[35] Because the township admits no responsibility for the chemical contamination, it has also felt no reluctance to force the residents to hook into the municipal water system. The township sees the toxic contamination as an unavoidable risk assumed by the victims when they bought their homes.[a] The *Asbury Park Press* reports, "Phillips (a Township Committeeman) said the residents should have been aware they were moving to homes that were close to the landfill. Most of the homes with the water problems are less than two years old. 'Still, they are part of the town and we have to do whatever we can for these people.'"[37]

Because the township does not assume responsibility for contamination of the aquifer, it has not cooperated with state attempts to close and seal the landfill. Although the township responded to the requests of Legler residents by trying in 1978 to close the landfill to septic tank sludges, it balked at closing the landfill to solid wastes and at sealing it from the environment. Jackson Township still received fees from disposal of solid wastes: $20 per forty-cubic-yard load. More importantly, however, it would be prohibitively expensive to seal the landfill permanently—perhaps over one million dollars.[38]

In January 1980, DEP took Jackson Township to court to force it to install replacement water supplies and to close the landfill permanently. As just noted, the township finally did complete the replacement water system two months after the deadline set in court and two full years after the groundwater was found to be contaminated. As of mid-1981 the township had yet to submit a plan which was acceptable to the state for permanently sealing the landfill. The township has proposed to cover the landfill with dirt and has phased out use of the landfill, but it will not take on the more extensive cleanup necessary to close the site. The township itself does not feel responsible either for the delays in the construction of the water system (as noted earlier, they had to wait until the extent of the contaminated area had been determined and until the plan received Pinelands Commission authorization before construction could start) or in closing the landfill. In fact, Martone feels the state needlessly delayed construction efforts.[39] The township has balked at the cost and effort of permanently cleaning, closing, and sealing the landfill both because it does not acknowledge the landfill as the source of the pollutants and because the township does not have adequate financial resources.[40]

[a]As of February 1982, no trial date had been set either for the victim's suit or for the township's counterclaim. Both parties are currently engaged in "discovery" proceedings.[36]

Three state agencies have been primarily responsible for responding to Jackson's contamination problem: BPU, DEP, and the Department of Health. Since the first public announcement of foul water in 1978, BPU has worked at cross-purposes from Health and DEP. BPU tried to keep the landfill open while the other agencies were trying to close it. As we have seen, BPU downplayed the dangers of toxic contamination from the landfill, assuming unless shown otherwise that the municipal landfill was not the source of the contaminants. While DEP did finally act to close the landfill, this agency and Health were also reluctant to assume that the landfill could cause a potential health disaster. According to both Legler residents and Jackson Township officials, the state did not respond to their early complaints about foul water and health problems. DEP's preliminary water sample findings showed that there were no dangerous levels of chemical contamination in the water; and the agency was uncooperative about releasing its later findings. Moreover, DEP's Solid Waste Administration certainly had been less than vigilant when it first granted the township its license to open the landfill on the Glidden-Durkee site.

It took the state more than a year to bring court action against the township to try to close and seal the landfill permanently and to get a replacement water system installed. DEP's top priority was to pressure the township to install a replacement water system.[41] Only after negotiations with township authorities failed did DEP turn its case over to the state's attorney general. DEP did not take the township to court for failing to meet its June 1, 1980, deadline to install the water supply; by that date the system was under construction. It was only on January 14, 1981, that DEP went back to court to force the township to comply with the earlier court order to close and seal the landfill.

The state's Health Department was also reluctant to assume that the toxic contamination in Jackson posed an immediate health risk. When the landfill was in operation, Health never considered the possibility that the huge quantities of septic wastes being dumped there might threaten public health in any other way than to encourage bacteria and pests, even after the township began to report illegal dumping of chemicals. When toxic wastes finally were found in the groundwater, the only act the Health Department initiated was to state that long-term exposure to these contaminants could cause chronic disease. When possible kidney disease was suggested by Gogats, Health found that "there is no common pathology in any two of the three kidney problems reported."[42] The department never considered the possibility of other immediate or short-term health effects from the contamination. The township's Board of Health pressured the state to conduct a health survey of the residents; but it was only after the Legler residents obtained a health survey schedule on their own from the Love Canal victims in New York, and administered it themselves, that the state analyzed the results.

Dhun Patel, the state Health Department's Chief of Environmental Health and Hazard Evaluation, noted that the chances of serious health effects to the Legler residents were very small since toxic wastes had been in the water supply for no more than six months before they were discovered; 40 percent of the Legler residents had moved to the area during the summer of 1978. Even if some residents had had longer exposure, according to Patel, it would be impossible given such a small population to quantify their exposure or to determine statistically what health problems among the population had been caused by the contamination.[43] Gogats noted that reports of foul water had been appearing since early 1977,[44] however. Health itself noted community health problems it had found in analyzing the victims' questionnaires:

> The repeated questioning about health effects is partially rooted in anxiety which has its cost in terms of emotional stress, too. While the community was drinking the water, there may have been a problem with diarrhea. Washing in the water is probably responsible for some skin irritation, but there is no indication of more serious health problems related to the water.[45]

Since September 1980, Health has been conducting a survey of the Legler residents in order to follow up on some complaints prevalent in the self-administered health survey, primarily skin problems, and in order to monitor the Legler population for possible long-term health effects such as cancer or birth defects. Immediate effects noted by the victims probably will not show up in the most recent survey, however, primarily because the victims have been drinking noncontaminated replacement water since December 1978.

The main impediment to effective response to the contamination in Jackson by either the township or the state authorities has been their lack of fiscal resources in contrast to the magnitude of the problem. In the late 1970s and early 1980s toxic contamination has been found in groundwater supplies all over New Jersey. Every community with contamination has requested the state Health Department to conduct a health study, DEP to monitor water supplies, and state enforcement officials to help discover the sources of the contamination. The state does not have enough resources to fill all these needs.[46] For example, a DEP court case against any single township (as with the Jackson case) requires scientific and health expertise, the time of state lawyers, and management by top DEP officials.

The township, as just noted, is faced with similar resource constraints. It has resisted pressure from the victims and from the state, and does not acknowledge responsibility for the contamination. The township was reluctant to install replacement water supplies or to seal the landfill because it did not want to spend money on behalf of a small number of its residents. The total cost of just replacing the water supplies to the victims was $1.2 million, around $48 for each resident of the entire township. Cleanup of the landfill would incur similar additional costs. Such costs are not easily met in a municipality already facing fiscal problems.

The Legler area residents have received little support from other Jackson residents, who resent having to pay higher taxes on behalf of this small group of pollution victims. Resentment has also been expressed toward the victims because the publicity about toxic contamination has hurt Jackson's overall housing market, one of the township's most important businesses. For example, a representative of the Jackson Chamber of Commerce wrote an editorial in the local paper suggesting that because only 165 homes out of over 8,000 were affected by contaminated groundwater, the victims should not be receiving so much attention from the township government.[47] Officials in Jackson may hold similar sentiments. Martone noted that the members of the township committee stem primarily from business and professional classes, those who have a stake in Jackson's growth. The township committees in the late 1970s and early 1980s included a young attorney, a housewife, a pharmacist, a small business executive, an engineer, a corporate accountant, a construction supervisor, a small business comptroller, a retired sales representative for IBM, and a sales representative for an oil company.[48]

Organizational impediments have also hampered effective action by either the township or the state. State and local officials, for financial or administrative reasons, have defined their functions narrowly. Whenever possible they have excluded full analysis of the dangers of toxic waste contamination in this area. The conflict between

DEP and BPU epitomizes this set of attitudes. Likewise, while the landfill was in operation the Health Department tested groundwater only for bacterial contamination even though both Health and DEP had been informed of possible toxic waste dumping by the township and local residents. Reasons why the two agencies ignored these obvious warning signals remain unclear,[49] although state officials may simply have believed that the landfill could not possibly be leaking since it was presumably located over a large layer of clay.[50] The state has also claimed that when foul water was first reported in 1975, it lacked the manpower to investigate the possibility of serious groundwater contamination. A final reason may explain the state's reluctance to investigate possible groundwater contamination: just as BPU had an institutional stake in keeping landfills open for their economic benefits, DEP and Health may have felt a stake in keeping as many landfills operating as possible, despite their localized hazards. The state's chemical wastes and septic tank sludges had to go somewhere.

The original decision to construct the Jackson landfill was made by default, as we have seen. Important information about the condition of the proposed site was never considered by the state regulatory agency. One of the reasons state officials seem to have neglected the landfill's potential health hazards is that the three state agencies which deal with landfills—BPU, DEP, and Health—are responsible for landfill siting and promotion as well as for landfill regulation. Landfills provide economic, health, and environmental benefits for the state as a whole, even if they produce disbenefits locally. This helps explain why, when the township's Health Officer asked DEP to investigate the original reports of foul water and pH changes, DEP officials responded that the Jackson landfill was "one of the best in the state."[51]

The most blatant examples of official neglect of potential hazards at the Jackson site seem to lie in the possibility that township officials were aware from the beginning that the soil under the landfill site was unsuitable, and that the township knew as early as 1975 that toxic wastes were indeed leaching from the landfill into nearby groundwater supplies. The boring logs are said to have misrepresented soil conditions. If the township knew about the leaching, it never made this information known to the appropriate state officials. One assumes that such neglect would have resulted from the township's vested economic and political interest in not making these dangers known. Local officials also have tried to avoid responsibility for the health and property effects of the landfill by pointing to the state's presumed responsibility for testing water supplies, conducting health studies, financing a replacement water system, and closing the landfill. Martone stated, for example, that the function of the township's Board of Health lies exclusively in preventive medicine; thus it could not play a large role in surveying victims' health. By pressuring the state in public, the township gave the appearance of trying to help the victims while it was actually passing the buck.

The Legler residents have played a considerable role in bringing the toxic contamination problem to light and in forcing state and township officials to respond to it. In 1971 local residents were the first to question the safety of the planned landfill—to no avail. In 1977, new residents rather than original members of the Concerned Citizens Committee began to complain about foul water and health problems. Once state tests indicated that well water was contaminated, however, the Concerned Citizens Committee was revived and began to represent the victims' interests in dealing with the township. As McCarthy relates, when the township first informed the residents in December 1978 that their water was contaminated, it refused to identify the contaminants, claiming that the problem was "too technical to explain."[52] (According to Gogats and Martone, this was the same response the township had received from the state.[53]) At this point, McCarthy and a neighbor hunted through state records looking for the source and nature of the contaminants.[54] The Legler residents were also responsible for discovering the adverse health effects of the contamination. At various meetings, residents compared their health problems, showing the rashes from taking showers and discussing more major illnesses, such as kidney disease. When the state Health Department refused to conduct a health survey, the residents obtained their own. They were also responsible for exposing the apparent misrepresentation of the landfill's soil conditions by Glidden-Durkee and the township engineer. Their blue-collar, working-class backgrounds certainly have not prevented the Legler residents from organizing effectively.[55]

Once the nature of the toxic contamination became known, the Concerned Citizens Committee took responsibility for pressing the state and township to take samples, replace the water supply, and close the landfill. When the township began talking of building water supplies, the Concerned Citizens organized to attempt to avoid having to pay for them. They sued the township, attempting to invalidate its mandatory hookup law. The citizens obtained an injunction forcing the township to maintain deliveries of water to residents not connected to the system. The *Asbury Park Press* reported, "Fifty-two Legler homeowners filed suit in May, claiming they were not responsible for the water pollution and should not have to pay to use the municipal water system."[56] The court upheld the township ordinance insofar as homeowners who live within 200 feet of the water mains would have to hook up to the system so as to not tear up roads in the future. In October 1980, ninety-six families filed a suit against the township and its officials, the twenty-one companies that had built houses in Legler, and two septic tank disposal firms for contaminating the groundwater, allowing it to be contaminated, or not informing the residents that it was contaminated. The victims sought $25 million for cleanup of the landfill and $1.5 million in property damages.[57]

The Legler residents have used the press and the polls, as well as the courts, to try to force substantive responses from local and state officials. As soon as contamination was found in the well water, the victims began criticizing DEP publicly for its delay in sampling and analyzing the water. According to McCarthy, press publicity was also instrumental in the victims' efforts to pressure DEP to sue the township to begin construction of the water system and to close the landfill permanently. For fifteen months the residents had requested DEP to put pressure on the township but had gotten very little response. As McCarthy sees it, only after the victims made DEP's lack of cooperation known to the press did the DEP file suit.[58] DEP's Onsdorf feels, however, that the actions of the victims had little to do with DEP's decision to take the township to court. To DEP, the victims are only "a very vocal interest group."[59] This doesn't explain, however, why it took DEP a year before the agency began demanding that the township seal the landfill and install water lines, nor why DEP responded rather quickly once its recalcitrance was exposed in the press. The residents also were successful in using public criticism to obtain a study from the state Health Department.

In 1979 the Legler residents, having gained no cooperation from the Democratic Party-controlled Township Committee, campaigned for Republican Party candidates who promised to help them. In November two incumbent committeemen and the mayor were defeated. As the *Asbury Park Press* reported:

> McKenna (the new mayor) said voters were dissatisfied with the way township officials handled water pollution problems in the Legler section. "The Township should be more responsive to the needs of the people," McKenna said. "The problem should have been handled quickly without six months of bickering about whose fault it was."[60]

Unfortunately, according to McCarthy, the new group of local officials, once in office, were little more responsive than their predecessors had been. In his blunt critique: "The two bums we replaced them with were worse than the ones we got out."[61]

Not all of the Legler residents see toxic waste contamination as a political issue to be contested between the residents and the township; a small proportion of them have chosen not to join the Concerned Citizens Committee. A few residents still drink their traditional well water; but, as McCarthy noted, the water can look, smell, and taste clean and still be contaminated.[62] There have also been pressures on the residents from outside, trying to discourage them from organizing against the state and township. When contamination was first found, the township persuaded about thirty Legler residents not to join the other residents in suing the township, promising that replacement water supplies would be free and available as soon as possible. Because New Jersey law states than any citizens suing a governmental body must inform the officials of their intention to sue within ninety days of the original infraction (in this case, the announcement that the groundwater was contaminated) or lose the right to sue,[63] these thirty citizens have been excluded from seeking subsequent recourse.

The legal precedents for damage cases also discourage suits such as the one brought by the Concerned Citizens. One standard environmental law textbook, for example, points out that

> A system of damage awards may fail by a large margin to reflect all of the losses associated with environmental degradation, particularly in the case of activities that create a risk of harm that may or may not eventuate in the future, or where it is highly difficult to attribute responsibility for a given damage, such as emphysema or cardiovascular disease, to a single cause.[64]

As the 1980 state Health Department report stated, the major health effects of the toxic contamination in Jackson may be the risk of future chronic disease. Because of the small population, even if cancer or birth defects result from the contaminants it will be difficult for the Legler residents to show a causal relationship.[65] For the immediate health effects from contamination—intestinal and skin disorders—to be actionable, the residents have to show both that they suffered these complaints and that this suffering was caused by the contaminants. "Under all of the traditional common law liability rules, plaintiff must establish a firm causal link between the conduct of defendant and actual or imminent damage to the plaintiff."[66] Because Health delayed its studies for nearly two years and because DEP has been reluctant to release data to the residents for a third party suit,[67] the residents' chances in court have been further reduced. Finally, both the state and the township are protected from citizen suits by the New Jersey Tort Claims Act. In addition to the ninety-day deadline for announcing intention to sue, this act stipulates that a governmental body cannot be sued for action or inaction on any law or governmental policy. In case a governing body is determined liable for damages, the plaintiffs are not entitled to damages for pain and suffering where medical expenses do not exceed $1,000.[68]

As has been noted above, Jackson Township put great pressure upon the Legler residents primarily for economic reasons. It filed a countersuit in October 1979; it passed the mandatory hookup law; it threatened those residents who did not connect with the municipal water system with fines, imprisonment, and curtailment of water deliveries; and it tried to force the residents to cap their wells although (or because) the residents were taking water samples from their wells to use as evidence in their lawsuit.

Other residents of the community have also tried to discourage the Legler victims from following up on the causes of water contamination. The editorial criticizing the residents has already been mentioned. McCarthy knows of four threats against his life, including two actual attempts to kill him.[69] As well as suing the township, the Legler residents are suing Glidden-Durkee for misrepresenting the geological conditions under the landfill site, and the victims' attorney has hired a private detective to investigate the original generators and carriers of the toxic wastes. As McCarthy sees it, private money, jobs, and possible criminal cases are all at stake.[70]

Despite both the procedural and substantive attempts to discourage their efforts, the Legler residents originally acted in 1978 because they wanted relief from health and property damage. In order to organize, however, the residents had to have seen the groundwater contamination as a common *political* issue. From the first indications of foul water, the Legler residents reported that they received little attention from state and local officials. The township's original announcement in 1978 that well water should not be consumed revived the Concerned Citizens Committee because the township refused to say what was contaminating the water or where the contaminants came from.

Beyond the issue of the actual contamination is the residents' feeling that they have been treated unfairly by the township, first when it decided to approve the landfill and in its subsequent actions. The township's original 1972 decision to ignore the wishes and safety of the residents because Glidden had offered the township officials "a deal we could hardly beat" has been representative of what the residents see as the state's and township's constant concern for financial factors over public health risks. As soon as toxic pollutants were discovered, the victims were convinced that the contaminants came from the landfill. A septic tank cleaning business, for example, which was located near the landfill received threatening phone calls and incurred vandalism soon after DEP began its investigation of groundwater contamination in 1978. The residents recognize their common opposition to the self-interest of the state and local officials for, as McCarthy states, as the residents face more resistance from officials they feel more reason to pursue their own demands.[71] This pressure has led to internal cohesion among the victims, as will be discussed later.

Much of the work of the Concerned Citizens Committee has been done by Jim McCarthy in consultation with the other members. McCarthy, like the other Legler residents, comes from a working-class background. Until the township proposed the Legler landfill in 1971, he was not very concerned either with politics or environmental quality.[72] In 1971, however, he became a founding member of the Concerned Citizens Committee. Attempts by this group to halt the landfill were rebuked by the township officials at that time and, as McCarthy relates, he left the town hall "with his tail between his legs."[73] In 1975, his nine-month-old daughter died of kidney cancer, contracted while still in the womb.[74] McCarthy was also subsequently disabled by kidney disorders which he attributes to the contaminated water. In comparison to most of the other Legler residents, however (over half of whom had moved to Jackson after 1975), McCarthy is an "old timer," having moved to his home in Jackson from New York City in 1969. Given his greater personal involvement in Jackson Township politics, his belief that the death of his daughter was the result of hazardous waste contamination, and the fact that he does not work because of his kidney disorders, McCarthy has been the most active of the Legler victims in seeking recourse for the group. As he related: "I look forward to the court case (against the township). I want to show people the injustices that have gone on in Jackson Township."[75]

McCarthy is not alone, however, in his attempts to gain recourse against the township. The Concerned Citizens Committee has a board of directors who supervises the group's attorney, Arnold Lakind, an environmental lawyer with the Lawrence, New Jersey, firm of Zauber, Szaferman, Lakind and Blumstein. Most of the issues dealt with by the group relate to the specifics of forcing the state, township, and courts to respond to the victims' demands for recourse.

To date, the victims have largely been ineffective in obtaining rapid, free replacement of the water supply, gaining timely health studies, closing the landfill, and receiving compensation for their wells. Although the Legler residents have obtained public responses to all of these demands, much time and effort have been expended in the process. Outside of using the press and the courts, this small group of residents has little political power, especially because they are in an adversarial relationship with their own local government. Improving the condition of the Legler residents will cost the state and township money and, therefore, is unpopular among many other township residents. Politicians choose carefully the issues they support so as not to alienate voters.[76] In the case of the Legler residents, while a new slate of officials was elected to township office in part because of their sympathy with the Legler residents' demands, once in office they realized that it was politically dangerous to support a small group of victims in opposition to the economic interests of many of their other constituents.

Because most of them have met with bureaucratic indifference, the Legler residents have rejected the possibility of dealing with the broader issues of toxic waste management in New Jersey, choosing instead the specific goal of obtaining a damage award. Jim McCarthy sees the experiences of the Legler residents as a struggle with both elected and career officials.[77] Thus, most Legler residents have not viewed either local or state government—except the courts—as a forum for the expression of public demands or interests. Even when they succeeded in replacing Democratic officials hostile to their efforts with Republican ones, the new Republican administration expressed platitudes in favor of the victims and then threatened them for not connecting to the municipal water system.

Because the Legler residents are in an adversarial relationship with both local and state government, they represent only their own localized interests. In such adversarial relationships, each set of actors musters as many political and legal resources as possible in an attempt to force a decision upon the other set of actors. This type of political conflict has focused the victims' attention on specific issues. When it came to general issues such as statewide toxic waste management, McCarthy explained, "People become gun-shy." McCarthy himself made it very clear in several interviews that he was not a "politician" or a political activist, because to be a political activist means to him that one has political ambitions.[78]

The Legler residents (aside from McCarthy) seem to have rejected dealing with broader issues of toxic waste management at least partly because, even if they receive a favorable ruling in court, they will not have had meaningful input into the policymaking process. When the township was ordered by the court in January 1980 to begin construction of the replacement water system, the victims' principal substantive demands were met. If trust funds are set up to clean up the landfill and meet the health needs of the victims, they will have influenced the local responses to contamination. But the broader issues of public responsibility to foresee and avert such health disasters in the future will not have been approached. Neither state nor township decisionmaking processes will have been opened up to effective public scrutiny. The suit does not question the fact that the township refused to acknowledge the original petitions of the Legler residents nor does it question the minimal responses of DEP and Health. The victims' suit does not question the township for having placed all responsibility for analyzing health effects, water quality, and landfill management on the state, nor does it address the issue of whether hazardous waste generators and transporters should be obligated to accept liability for whatever health and property damage they have caused. In fact, the procedure for a citizen suit against a government body exacerbates the tendencies of public officials to be inaccessible to citizen scrutiny and requests. As already noted, public officials are not liable for having failed to fulfill their function. In order to gain recourse, citizens must show malfeasance and negligence. Thus, to protect themselves, public officials are likely to make data and decisionmaking records even less available to citizen scrutiny. McCarthy and the Concerned Citizens do admit, however, that their situation is not unique. In Plumsted Township, bordering Jackson to the east, DEP has been cleaning up seven abandoned toxic waste dumps. The Concerned Citizens have also had contact with the Love Canal Homeowners' Association.

The Legler residents have proved adept at using all of the political tools available to them—from the courts to the state DEP to the township elections. Only Jim McCarthy, however, has publicly raised the more general issue of improved statewide procedures for siting and managing hazardous waste facilities. Given his experiences as a victim, McCarthy sees a need in New Jersey for additional safe disposal facilities and sites where companies can dispose of their wastes relatively inexpensively in order to avert illegal dumping. He worked with New Jersey Senator Frank Dodd on a State Senate bill (S.1300) that establishes a state hazardous waste siting mechanism (see Appendix B). McCarthy gives strong support to the provision in S.1300 that calls for public participation in the siting process, and also feels there should be a half-mile buffer zone between any hazardous waste facility and the nearest residences.[79]

Thus McCarthy has translated his experiences as a victim to the more general issue of toxic waste management. He has given numerous lectures, speaking about his experiences as a resident. He feels that "what happened out here should never happen to anyone else and I hope to see that it never does."[80] Insofar as he has addressed questions of justice and bureaucratic responsibility to the needs of Legler victims, McCarthy's great personal investment into the particulars of the Jackson case has bearing on more general issues of hazardous waste management. He has taken on the work of representing the victims as a full-time job. He also feels strongly about the effects toxic contamination has had on the victims, having lost a child to what he feels was contaminated drinking water. McCarthy was also the first to recognize the connection between the township's decision to construct the landfill in 1971 and the government responses to toxic contamination in 1978.

At first glance, the moral of the Legler story seems clear: tremendous amounts of pressure must be applied to get a government body at any level to respond substantially to a set of local demands. The Legler residents feel that it has only been their cajoling of DEP and Health, and their use of the courts against the township, that have led to any response to their demands. The state has found the township unwilling to make the necessary investments to clean up the landfill or to run water lines to the victims. Likewise, the township has threatened court action against the victims, and has publicly chastized the state for not making a greater commitment to the health and welfare of the victims.

Some of the reasons for all this confrontation have been explored already. None of the actors has been willing to accept responsibility for the contamination; they do not want to make the politically unpopular and expensive move of trying to remedy the victims' situation. State and local officials have found it easier to react to crises such as the Legler contamination rather than taking the responsibility for averting possible future health dangers. Thus, the responses made by different officials vary in their sense of urgency. Officials have been shown to have different economic stakes in the provision of services to the residents. These officials had different information available, which they could choose to use or not at their discretion. There were also various levels of pressure upon the decisionmakers, all leading to the tendency of the state and township to try to commit as few resources as possible to the needs of the residents.

Finally, the lack of cooperation between any of the actors in the Jackson case may be attributed to the biased perceptions each held toward the other's intentions. The Legler residents saw the actions of state and local officials as attempts to deny the residents their legally justified rights. Thus, the officials were thought of as inept, corrupt, or simply having "political" ambitions, based on what was politically most popular and economically most lucrative for the government. McCarthy has suggested that powerful figures in the state and in the township were in collusion to deny the residents their legal rights.[81]

Yet, given the conflicts apparent between the township and the state, this view of a monolithic governmental force seems incorrect. The township sees itself as having gone out of its way to help the victims, while receiving nothing but bad publicity in return for its efforts. The township claims that while it has a genuine interest in the welfare of the Legler victims, their needs must be balanced against the financial resources of the overall community. In the township's eyes, the Legler residents are making unjust claims. The township also sees the state as making unreasonable demands upon local economic resources, and of assuming that the township is guilty until it can be proven innocent. The state agrees that the victims are making unreasonable demands. To state officials, their agencies can only make decisions on the basis of available scientific information. The state sees the township as attempting to escape from its obligations to the welfare of its own residents. Because each actor feels that the others are trying to take advantage of the situation, little room exists for compromise, and conflict remains rancorous.

Even though health and property damage from toxic contamination may be very high and government assistance is often conspicuously absent, victims do not necessarily organize collectively to express their demands. Not only must they be aware of chemical hazards to their lives, but they must be able to assign responsibility for this contamination to specific actors and feel that their demands are legitimate within the existing political and social structure. Norie Huddle and Michael Reich describe, for example, the responses of victims of PCB poisoning in Japan. Although victims suffered boils, fatigue, weight loss, and birth defects,

> attempts by the victims to escape the "coldhearted eyes of others" contributed to delaying recognition of the disease and clarification of its origin. Indeed, victims were so successful in isolating themselves from both society and each other that, until October, many families believed that they alone were stricken.[82]

In the Jackson case, victims did not begin organizing until late 1978, a year and a half after their drinking water was first reported foul. What, then, leads victims to consider their plight to be a political rather than a personal problem? And to what extent have victims been able to extend these issues to gain support from nonvictims?

Probably the most striking characteristic of the Jackson victims is their solidarity. Even though McCarthy has done most of the research, organizing, and publicity for the group, he emphasizes that he does everything with the approval of the Concerned Citizens Committee. Nearly all of the victims have joined the Concerned Citizens. The group's unity is especially striking given the differing economic concerns of different sets of victims. Only 90 of the 165 families, for example, have joined the lawsuit against the township (a costly action). As noted earlier, some thirty families were persuaded by township officials soon after toxic contamination was announced that they would receive free replacement water supplies and thus should not join the lawsuit. Legler residents also differed in their response to the township's order to hook up to the municipal water system once it was constructed. Most families paid the hookup fee, some are still using private deep wells, and others have not connected to the municipal system out of indigence or principle. Despite these differing interests and legal stance toward the township, the victims have worked together. McCarthy states that the group as a whole sought the court order halting the township's plans to cease water deliveries. The Legler residents acted in concert when they campaigned against the reelection of township officials and when they challenged the mandatory hookup law. In other words, despite individual differences in material and legal situations, the Legler residents were united in their resistance, both ideological and material, to the state and township.

The Legler residents may have become such a unified political force in part because, while they are *not* unified by specific economic interests, they share a credible ideology and common symbols. Unlike many environmental groups, the Legler residents tend to be of lower-middle or working-class backgrounds. A large proportion of the victims, somewhat surprisingly, have little personal stake in the community; they are newcomers to Legler. As the New Jersey Department of Health related, "Nearly half of the population has moved in since January, 1978, and a third lived in the area for less than six months before the potable water supply was changed."[83]

The Jackson victims' ability to create a strong organization seems to be related to two facts. First, the victims contrast their own political and social values with those of the entrenched Jackson elite, the upper-middle-class businessmen and professionals who have an inherent interest in limiting publicity about the case and in resisting the victims' demands. When chemical contamination was first identified in 1978, McCarthy and the other original members of the Concerned Citizens Committee unified their neighborhood around symbols of health threats and of government indifference, saying essentially that the victims had to join together to defend their interests against those of both the township and state governments. To support their claims, the organizers of the resuscitated committee could point both to the township's original decision to site the landfill in Legler and to the lack of response by state and township officials to the initial reports of contaminated water. As the nature of the chemicals became known and the victims met more resistance from township and state officials, not only did the legitimacy of the victims' demands grow but the original antigovernmental ideology around which they had organized was reinforced as well.

The Legler victims also have a strongly unified organization because they have discovered common political interests. The victims first discovered the health effects of the toxic contamination by comparing their personal health histories and by conducting the first health survey in the area. They also were the first to report the original connection between Glidden-Durkee and the township which had led to use of an unsafe landfill site. Finally, the more effort the victims put into the work of the committee, naturally the closer their interests have been allied.

Love Canal victims, in contrast, have expressed little community of interest. Since its inception, the Love Canal Homeowners' Association has been fraught with internal dissent. Racism has divided members of the neighborhood. Divorce or separation have rent 40 percent of the Love Canal families. Most of these conflicts can be traced to economic causes: Love Canal victims are competing (as they see it) for limited public relief funds. Michael Brown writes,

> Bitter arguments also divided the Love Canal Homeowners' Association. Those who lived immediately near the chemicals believed there would be only a limited amount of government money forthcoming, and they demanded that they be the first to go. . . to fight for the largest portion of the funding they formed a radical splinter group.[84]

Even divorces have been traced in part to essentially economic conflicts. Husbands have resisted moving out of their contaminated homes at less than market value, while wives expressed more fear about the effects the chemicals were having on their families.

These victims' failure to cooperate does not seem to stem from a lack of knowledge of their plight. The Love Canal residents recognize the types of health and property damage they have incurred, and also the fact that local, state, and federal officials have made minimal efforts to alleviate the situation. The legitimacy of the victims' demands grows in the victims' own eyes with the discovery of each new chemical, each revelation of adverse health effects, identification of each new set of victims, and as official promises fail to hold true. Thus, when the New York State Health Department ordered the evacuation of pregnant women and children under two years old, all of the Love Canal victims were in the streets crying, shouting, and demanding to be moved. And with the discovery of dioxin leaking out of the canal, residents became violent, disrupting construction that they felt was spreading traces of dioxin around the area. Vandalism occured at the construction site, and seventeen residents were arrested.

Except in those few cases where Love Canal residents have directed their frustrations toward state officials or at the construction site, they have not been able to organize themselves to voice their demands effectively. Unlike the Jackson case where responsibility for the contamination could be attributed to the township itself, culpability and responsibility were not as easy to ascertain in the Love Canal case. No one actor could be identified as the sole source of the residents' complaints. The chemical wastes, of course, were dumped in the canal by the Hooker Chemical Company. But the company probably was not legally liable for the contamination, and furthermore was the source of jobs and taxes necessary for the community. Although the city owned Love Canal at the time of the controversy, it also could not be held fully responsible for the contamination. And while the state and federal governments were being asked to help the victims financially, they were not responsible for the initial disaster.

Because political organizers in Love Canal were unable to direct frustrations against specific actors, the victims felt themselves to be at the mercy of the goodwill of state, local, and federal officials. As a result, although their complaints were personally legitimate, they were not politically legitimate. The *New York Times* explained:

> Disaster victims traditionally have difficulty dealing with government agencies. . . but the difficulty seemed acute in the case of these lower-middle-class families: "they are good citizens. They pay their bills and their taxes and have served in the military. 'Why' they ask, 'shouldn't we be taken care of now?' "[85]

Health problems among Love Canal residents were also not recognized as having a common cause until the state Health Department ordered the evacuation of pregnant mothers and young children. Brown writes:

> The Schroeders looked upon these health problems, as well as certain illnesses among their other children, as acts of capricious genes, a vicious quirk of nature. Like Aileen and Edwin Voorhees, they were mainly aware that the chemicals were damaging their property.[86]

Unlike the Jackson victims, the original discovery of common health problems was not made by the victims themselves but rather by government researchers.

Because most of the Love Canal victims do not have a common political ideology which directs blame against specific actors, and because they are dependent upon government assistance as a concession to their plight, individual goals ceased to be political—that is, communal—but rather become economic. Individuals were concerned about whether they would have a chance to sell their houses.

In this light it is understandable why there was conflict among the victims in Love Canal. They tended to view their plight as a personal, economic matter rather than a common political issue; yet at the same time, the victims were not in control of their own economic and physical well-being.

> Along with the sense of abandonment and isolation, many in Love Canal feel a loss of control over their lives. "They realize that this loss of control stems from long-ago decisions to bury chemicals and then to build homes near that spot, not from decisions they made," Dr. Levine said.[87]

Furthermore, when it became clear that the state was going to purchase houses in the Love Canal area, it was unclear how many houses the state was to buy and where these houses would be. This uncertainty contributed to the victims' alienation. Brown writes,

> The ever-deepening official uncertainty and disorganization in developing a rescue plan had a marked effect on the attitude of the people and on their psychological stability. Unable to cope with jobs, men remained at home. Children, exposed to an inundation of television reports of radio messages on their danger, began to have nightmares of death.[88]

As a result of this uncertainty, conflicts arose within the community. Groups of victims felt themselves in direct competition for state monies rather than focusing their resentment on actors outside of the neighborhood. Furthermore, the feelings by victims that they had no control over their lives themselves led to frustrations that expressed themselves in violence. The *New York Times* summarized the views of one resident:

> It was a terrible time. . . . I didn't know if my baby was going to be born normal. I didn't know if the state was going to buy our house or if we would have to file for bankruptcy. So I struck out at my 5-year old. I physically abused her once.[89]

Because Love Canal residents felt they had little control over their own lives, they did not share a political ideology which explained the causes of their situation and which translated this knowledge into political action.

Not all of the Love Canal victims felt themselves alienated from one another and wrapped up in their own economic problems. The members of the Love Canal Homeowners' Association have been described as acting in a very unified fashion.[90] Furthermore, Love Canal residents have placed some organized pressure upon state and local officials, such as in the boycott of EPA health tests and the work of the Love Canal Homeowners' Association in locating new victims.[91] Yet the perception of common political values necessary for common political action has arisen only in cases where political demands have been directed toward actors *outside* of the Niagara Falls community. After the state refused to purchase more than 200 homes in Love Canal, for example,

> Hundreds of families in the polluted Love Canal neighborhood said today that they would "not allow" the EPA to conduct large-scale medical tests in the area, unless the government agreed to purchase their homes.[92]

This was the first time that six Love Canal groups—including homeowners, renters, church and synagogue members—had worked in concert with each other.

In order to pressure concessions from resistant government officials, victim groups have to mobilize public support outside of the contaminated area. Because they represent such a small constituency in relation to the amount of resources necessary to purchase houses, replace water supplies, perform health tests, and clean up contamination, such support for their demands is essential. In the Jackson case, the Legler residents interpreted the state's decision to sue the township as resulting from public agitation. Certainly New York's decisions to allow Love Canal victims to remain in hotels due to the dioxin threat and its eventual decision to purchase more houses were the result of the state's concern for its public image. Brown writes,

> Under this tension, the state had no choice but to allow the people to remain in the hotels until the construction work was finished. . . . They were in hotels under temporary provisions, more and more people vowed they would not return home. Finally, Governor Carey, reacting to the daily pressure, agreed that the state would move 200 to 550 additional families during the next two years.[93]

The victims' primary appeal to nonvictims involves the relative legitimacy of the victims' claims as opposed to those of public officials. Toxic waste victims essentially question the justice of limited public responses. Victims appeal to the fact that they have been victimized by the contamination followed by government insensitivity and unresponsiveness as well. The emphasis of these appeals is to make the victims' plight visible so as to arouse widespread sympathy.

Brown relates the criticism he received from public officials and the Hooker Chemical Company because of a series of articles describing what he saw as the health problems of the Love Canal residents. McCarthy and the Jackson victims have also been the subjects of extensive media coverage. The Channel 7 (New York) program, "Poison Water," aired on November 19, 1980, emphasized the attempts by victims to arouse sympathy among the general public.

> The half-hour show consisted almost entirely of interviews with Jackson residents—in various poses of pain, anguish or anger, with music added in the background—discussing their health problems and blaming them on the town's polluted groundwater supply. Complex questions like the town's liability, how the water was polluted, why this problem might be occurring elsewhere and what implications it has for public policy were never discussed.[94]

In order to appeal to public opinion, victims often attempt to manipulate media symbols. Their primary message is that they have been victimized. Therefore, they not only show the effects of their victimization but may also try to make actual acts of victimization visible to the public. For this reason, they may engage in civil disobedience. Jackson residents defied the mandatory hookup law and took the township to court for trying to cut off water deliveries. Love Canal residents demonstrated, were arrested, held public officials hostage, and refused to vacate hotel rooms. Any retaliation by government against the victims simply reinforces the public image which the victims are trying to create.

> The other day Mrs. Gibbs and a group of mothers marched into a Niagara County legislators' meeting, wearing red carnations as "the hostages of Love Canal." The mothers then threatened to hold the legislators hostage until they took some action on a proposal concerning the canal. After the board acted, Mrs. Gibbs told a reporter: "We didn't have any guns or anything. We were just planning to body barricade the doors."[95]

One of the reasons victims can raise sympathy among the general public is that many of the dangers of hazardous chemicals are now common knowledge. Because of high media attention to hazardous chemicals, nonvictims may vicariously experience the plight of victims through concern for their own health. Chemical drums, midnight dumping, and "Love Canal" are a set of symbols which embody public fear of chemicals. At the same time, this public fear can backfire in such a way that the victims themselves may be associated with the chemicals. As a *New York Times* article reported about Love Canal,

> Dr. Levine cites the case of the woman who liked to go to a downtown bar for an occasional drink. No matter which end of the bar she stood at, everyone moved to the opposite end.[96]

Huddle and Reich described the social stigma attached to chemical poisoning victims in Minimata, Japan:

> Believing the disease to be contagious, [the public] cautiously avoided and eventually ostracized the afflicted. Shopkeepers refused to serve them or their families, and when the victims passed along the street, their awkward gaits and physical deformities brought stares, smirks and harsh ridicule from onlookers.[97]

Yet where victims attempt to avoid the "coldhearted eyes of others," they will be unable to mobilize any public attention.

Public demands by victims inherently question the justice and legitimacy of public officials. When victims attempt to influence public opinion, they are claiming that their health and property have not been properly protected. Because, however, the stated function of public officials is to protect the public welfare and because officials are politically vulnerable to these charges, they will attempt to deflect or refute the charges, showing that victims' claims are unfounded or appealing to other values held by the public. In Jackson, officials tried to silence the victims essentially by making their lives more difficult. Thus the township filed a countersuit, threatened residents who didn't connect to the water supply with fines and imprisonment, attempted to halt water deliveries, and passed a well-capping ordinance. At the same time, township officials claim to have fulfilled their humanitarian obligation to the victims through the delivery of water in tanks for two years and the construction of a new water system. To the township, the Legler victims are ingrates, suing the township despite the efforts it has made on their behalf. Public officials' statements made after the 1979 Jackson Township election calling for greater responsiveness "to the needs of the victims" can also be seen as attempts to convince nonvictims that the officials do have the victims' interests in mind. Township officials and local leaders also appealed to Jackson residents against the demands of the victims by referring to the expense entailed in meeting these demands. The Jackson newspaper carried several editorials by Chamber of Commerce members pointing to the economic effects the victims were having on the community as a whole.

Similar appeals seem to have been made by Niagara Falls officials. Local and state officials resisted publicizing the health effects which the contamination had on the victims, and also resisted the release of reports that new chemicals had been found. At the same time there seem to have been appeals to the economic dependency of Niagara Falls upon Hooker Chemical Company and concern over the high economic demands of the victims. The *New York Times* writes: "From the outside, [the Love Canal victims] were sometimes seen as 'making a fuss' over a few chemicals or 'making a killing' from the state on their homes."[98]

The effectiveness of victims' appeals to nonvictims thus depends in large measure on the extent to which nonvictims identify and empathize with the plight and demands of the victims. In Jackson, the Legler residents have very little local support because, if for no other reason, they are suing the township for sums that will bankrupt not only the municipality but, in the last analysis, Jackson's other taxpayers. Most of the Jackson victims' appeals to public opinion have been directed toward New Jersey as a whole. And the Jackson victims themselves are unified. In Niagara Falls, the Love Canal victims have not made exorbitant demands upon local government. Rather, they have received assistance primarily from the state and the federal government. Because the costs of relocating the victims are diffused across the broader political system, the Love Canal Homeowners' Association and other groups seem to have been more successful at appealing to the sympathy of Niagara residents as well as to residents of the state as a whole, even though—ironically—the victims themselves have a far less coherent organization.

CHRONOLOGY OF EVENTS IN JACKSON TOWNSHIP

1970

First landfill closed because of residents' complaints.

1971

Glidden offers the township mined-out land.

Legler residents sign a petition against the new landfill.

Township receives and submits to the state inaccurate soil-boring logs.

State Solid Waste Administration (SWA) and Board of Public Utilities (BPU) approve landfill.

1972–1976

300,000 gallons of septic wastes brought to site per day.

State cites township for ponding of septic wastes.

1975

Township became aware of toxic chemicals in landfill (according to state suit).

1976

Township petitions BPU to reduce delivery of septic wastes to 20,000 gallons per day.

Chemical fire at landfill.

March 1977

Residents first report finding their well water foul and dangerous to their health.

Township Health Department begins investigation.

November 1978

DEP begins investigating report that water is foul.

Residents told to avoid drinking their well water.

At residents' request, township closes landfill to further sludge disposal.

BPU orders landfill reopened for sludge disposal.

DEP and EPA criticized for lack of commitment to sampling and laboratory analysis.

DEP tests show no significant level of biological or chemical contamination.

Department of Health finds no common liver pathology.

December 1978

DEP finds organic chemicals in well water.

BPU refuses to close landfill until proof that chemicals are coming from septic waste dumping.

Township health officer chastizes DEP for slowness in analyzing samples.

DEP declares public health emergency.

Replacement water deliveries begin.

Concerned Citizens protest having to pay for new water system.

January 1979

Township asks state for relief money to construct water system.

Ocean County representatives in state legislature propose bill authorizing $1.2 million, 5 percent loan to construct water system.

May 1979

Township passes mandatory hookup law.

Residents respond with lawsuits.

July 1979

DEP finishes study of the size of contaminated area.

Study criticized by township health officer.

October 1979

Superior Court judge upholds hookup law.

Jackson passes $650,000 bond ordinance to finance water system; accepts bid on system.

Governor Brendan Byrne approves state loan to township.

Township files countersuit against residents.

November 1979

Township election: GOP candidates unseat Democrats on Legler issue.

DEP asks township to close landfill to all wastes and build cover.

January 1980

Legler residents finish health survey and send it to Health Department for analysis.

Responding to DEP court action, judge orders township to construct water system, close landfill to all wastes, and prepare plan to seal landfill.

February 1980

Township begins construction of water system.

June 1980

Court ordered deadline for completion of water system.

July 1980

Water system in place.

Health Department finishes analysis of Legler health survey.

Study is criticized by residents.

August 1980

Ocean County Health Department begins distributing new questionnaires to Legler residents.

September 1980

Legler residents file suit against Glidden-Durkee.

Township passes resolution demanding that residents hook into municipal water system as required under 1979 ordinance.

November 1980

Township threatens to stop delivery of water; residents obtain restraining order.

December 1980

Township law forces residents to cap their wells.

NOTES

1. Joseph Martone, Jackson Township Attorney, Telephone interview with David Moldenhauer, 13 January 1981.
2. Jackson Township Planning Board, *A Comprehensive Development Plan* (1964). It was only in 1974 when Great Adventure, an amusement park, moved to Jackson that this financial situation was alleviated.

Great Adventure is the largest taxpayer in the township, paying roughly a quarter million dollars per year. David Miller, Assistant Township Administrator, Telephone interview with Moldenhauer, 23 March 1981.

3. *Asbury Park Press*, 28 December 1978. Up until this action, municipal water supplies had been limited to the most densely populated section of the township, the only section where municipal water could be supplied economically. The costs of the new system in 1970 were borne by the residents themselves, all of whom were required by the township and state to tie into the water system. Miller, ibid.
4. Miller, Ibid.
5. *Asbury Park Press*, 29 December 1978.
6. Martone, Personal interview with Moldenhauer, 21 November 1980.
7. View provided by Keith Onsdorf, DEP Division of Enforcement, Personal interview with Moldenhauer, 15 December 1980.
8. Robert Gogats, Township Health Officer, Telephone interview with Moldenhauer, 13 January 1981.
9. *Asbury Park Press*, 29 December 1978.
10. James McCarthy, Spokesman for Concerned Citizens Committee, Telephone interview with Moldenhauer, 26 November 1980.
11. Jeffrey Hoffman and William Grey, *A Study of the Jackson Township Sanitary Landfill* (Princeton: Princeton University Water Resources Program, 1980), p. 10.
12. Ibid.
13. *Warren J. Adelung et al. vs. the Township of Jackson,* U.S. District Court Civil 79; 2613.
14. Gogats, Interview, 13 January 1981.
15. *New York Times*, 7 February 1980.
16. Gogats, Interview, 13 January 1981.
17. *Asbury Park Press*, 9 November 1978.
18. *Asbury Park Press*, 7 December 1978.
19. Gogats, Interview, 13 January 1981.
20. Hoffman and Grey, *A Study of the Jackson Township Sanitary Landfill,* p. 21.
21. McCarthy, Telephone interview, 29 November 1980.
22. *Asbury Park Press*, 22 July 1979.
23. Miller, Interview, 23 March 1981.
24. Martone, Interview, 13 January 1981.
25. McCarthy, Interview, 29 November 1980.
26. *Asbury Park Press*, 14 December 1978.
27. McCarthy, Telephone interview, 13 January 1981.
28. Martone, Interview, 13 January 1981.
29. Ibid.

30. *Adelung v. Jackson.*
31. Martone, Interview, 21 November 1980.
32. *Adelung v. Jackson.*
33. *Asbury Park Press*, 30 October 1979.
34. Martone, Interview, 13 January 1981.
35. Arnold Lakind, Attorney for the Concerned Citizens Committee, Telephone interview with Moldenhauer, 11 January 1981.
36. Telephone conversation with victim's law firm of Fauber, Szaferman, Lakind and Blumstein, 15 February 1982.
37. *Asbury Park Press*, 26 November 1978.
38. Martone, Interview, 13 January 1981.
39. Ibid.
40. Martone, Interview, 21 November 1980.
41. Onsdorf, Interview, 15 December 1980.
42. *Asbury Park Press*, 30 November 1978.
43. Dhun Patel, State Health Department, Personal interview with Moldenhauer, 15 December 1980.
44. Gogats, Interview, 13 January 1981.
45. New Jersey Department of Health, Division of Epidemiology and Disease Control, *Groundwater Contamination and Possible Health Effects in Jackson Township, New Jersey* (Trenton: New Jersey Department of Health, 1980), p. 8.
46. Onsdorf and Patel, Interview, 15 December 1980.
47. McCarthy, Interview, 26 November 1980.
48. Martone, Interview, 13 January 1981.
49. Onsdorf and Patel, Interview, 15 December 1980.
50. Gogats, Interview, 13 January 1981.
51. Ibid.
52. McCarthy, Interview, 26 November 1980.
53. Gogats and Martone, Interview, 13 January 1981.
54. McCarthy, Interview, 26 November 1980.
55. For another example of citizen organization, see Grace Singer, "People and Petrochemicals," in David Morell and Grace Singer, eds., *Refining the Waterfront* (Cambridge, Mass.: Oelgeschlager, Gunn, & Hain, 1980), pp. 19–63.
56. *Asbury Park Press*, 10 October 1979.
57. *Adelung v. Jackson.*
58. McCarthy, Interview, 26 November 1980.
59. Onsdorf, Interview, 15 December 1980.
60. *Asbury Park Press*, 7 November 1979.
61. McCarthy, Telephone interview with Moldenhauer, 29 November 1980.
62. Ibid.
63. New Jersey Tort Claims Act, Public Law, Title 59.

64. Edward Stewart and James Krier, *Environmental Law and Policy* (Indianapolis: Bobbs Merrill, 1971), p. 201.
65. New Jersey Department of Health, *Groundwater Contamination in Jackson*, p. 8.
66. Stewart and Krier, *Environmental Law and Policy*, p. 112.
67. McCarthy, Personal communication with David Moldenhauer, 10 January 1981.
68. New Jersey Tort Claims Act, Public Law, Title 59.
69. McCarthy, Interview, 26 November 1980.
70. Ibid.
71. Ibid.
72. McCarthy, Interview, 29 November 1980.
73. Ibid.
74. *New York Times*, 7 February 1980.
75. McCarthy, Interview, 13 January 1981.
76. Robert Dahl, *Who Governs?* (New Haven: Yale University Press, 1961), pp. 197–199.
77. Various interviews with McCarthy.
78. McCarthy, Interview, 29 November 1980.
79. Ibid.
80. Ibid.
81. Ibid.
82. Norie Huddle and Michael Reich, *Island of Dreams* (New York: Autumn Press, 1975), p. 138.
83. New Jersey Department of Health, *Groundwater Contamination in Jackson*, p. 5.
84. Michael Brown, *Laying Waste* (New York: Pocket Books, 1981), p. 36.
85. *New York Times*, 16 May 1980.
86. Brown, *Laying Waste*, p. 7.
87. *New York Times*, 16 May 1980.
88. Brown, *Laying Waste*, p. 36.
89. *New York Times*, 16 May 1980.
90. *New York Times*, 26 May 1980.
91. Brown, *Laying Waste*, p. 46.
92. *New York Times*, 21 May 1980.
93. Brown, *Laying Waste*, pp. 55, 57.
94. Bill Schmitt, "Environmental News in New Jersey: Beyond the 'Time-bomb,' " (Paper for Engineering 303, Princeton University, January 1981).
95. *New York Times*, 16 May 1980.
96. Ibid.
97. Huddle and Reich, *Island of Dreams*, p. 95.
98. *New York Times*, 16 May 1980.

APPENDIX B

AN ANALYSIS OF NEW JERSEY'S SITING SOLUTION

The purpose of New Jersey's "Major Hazardous Waste Facilities Siting Act" enacted in 1981 is to ensure that the state will have the ability, despite possible local opposition, to establish sufficient waste disposal capacity to meet its needs. Without the additional capacity offered by "environmentally acceptable" facilities, many fear that extensive waste disposal will take place illegally, to the detriment of the state's environment and its residents' health.[1] In the final hearings in the state senate on this legislation, the consensus of most experts was that three to five new waste management facilities would be needed to handle New Jersey's hazardous waste requirements adequately.[2]

PASSAGE OF SITING LEGISLATION

Senate Bill 1300 (S.1300) was introduced into the state legislature in 1980 as a way to cope with local opposition to the siting of hazardous waste facilities in the state. This opposition was quite evident, for example, in the rejection of a proposed landfill in Bordentown, detailed in Chapter 2. The final version of this bill, entirely rewritten over the course of fifteen months, was signed into law on September 9, 1981 (Chapter 279, Public Law of 1981). This statute combines state preemption with procedures for public participation and local compensation, and establishes new institutions and mechanisms for locating hazardous waste facilities.

The original version of S.1300 had been introduced not long after Governor Brendan Byrne's Hazardous Waste Advisory Commission had issued its recommendations in January 1980. Executive Order 76 in August 1979 had authorized an ad hoc group to study ways to improve the state's hazardous waste management program, with special attention to evaluating alternate methods the state might employ to assure future siting of hazardous waste facilities. The principal recommendation of this group was that a new corporation be established which would be responsible for planning, preparing, and managing an overall hazardous waste program for New Jersey.

This corporation concept was soon embodied in the form of legislation drafted by the Department of Environmental Protection (DEP) and introduced into the senate by Senator Frank Dodd on June 9, 1980. Objections to this bill quickly surfaced. They rested on both procedural and substantive grounds. Some felt that the bill had been introduced before the concept of a management corporation had received sufficient scrutiny by the public. Indeed, the Governor's advisory commission had recommended creation of an interim board to conduct further analysis of institutional options, as the basis for an informed decision on the preferred institutional structure. Instead, DEP and Dodd proceeded with the concept of a new state corporation.

More importantly, there was a widespread perception that the original bill failed to institutionalize properly the public's concerns about siting, particularly concerns of citizens in communities where proposed facilities would be located. Required public hearings were the main avenue for public participation in the bill as originally drafted. Moreover, local representatives were concerned that the corporation could exercise eminent domain power to obtain sites, with few constraints. Industrial groups also objected to the bill, in part because it allowed for the operation of facilities by the government, a form of publicly subsidized competition with private disposal operations.

As a result of these widespread objections, Senator Dodd, chairman of the Senate Energy and Environment Committee, agreed to an unprecedented process of restructuring the bill. This revision entailed not only a series of committee hearings and public hearings on the bill, but the establishment of an informal negotiation process that included four major interest groups: the chemical industry, the hazardous waste disposal industry, environmental and public interest groups, and municipal officials. In January 1981, this rewritten S.1300—now a product of negotiation and compromise—was adopted by the senate committee (as S.C.S.1300) and approved by the senate. Review of this legislation by the corresponding assembly committee resulted in far fewer revisions. Approval of the bill by the assembly on June 25, 1981 was unanimous (59-0),[3] and Governor Byrne signed the bill into law in September 1981.

THE S.1300 SITING PROCESS

New Jersey's siting legislation creates new bodies: a Hazardous Waste Facilities Siting Commission ("siting commission") and a Hazardous Waste Advisory Council ("advisory council"). In conjunction with the state's regulatory agency (DEP), these two bodies are responsible for carrying out the legislation's siting procedures.

The siting commission consists of nine members appointed by the governor. Its membership is intended to provide a balance of viewpoints: three members are county or municipal officials, three are from industrial firms, and three are representatives of environmental or public interest groups. The bill also provides for appointment of two temporary members when the siting commission makes decisions on specific sites. One of these temporary members is to be appointed by the governing body of the affected county, and the other by the governing body of the municipality in which the site actually is located.

The siting process begins with the premise that waste facilities must be located in those areas of the state that are the least "environmentally sensitive." Toward this end, the bill provides for DEP to adopt formal siting criteria.[a] Two sets of public hearings on these siting

[a]Five specific siting restrictions are set forth in the bill. These include a prohibition against siting a waste facility within 2000 feet of most occupied buildings. An interesting feature of the fifth restriction is its unmistakable political content: location or operation of hazardous waste facilities is prohibited: ". . . within a 20 mile radius of a nuclear fission power plant at which spent nuclear fuel rods are stored inside." This exemption is unlike the others in that it has virtually no bearing on whether or not a facility would be located in an "environmentally sensitive" area. Indeed, the prior siting of the nuclear power plant would tend to suggest an even greater level of environmental suitability. This provision was surely based on feelings of political equity. Those facing the risks of nearby nuclear operations are averse to encountering the additional risks of chemical disposal. This demonstrates quite clearly the contention that, despite their attention to technical factors, siting criteria have a strong political basis (see Chapter 5). The political aspects of siting criteria were also evident in the fact that the Assembly committee responsible for reviewing the bill passed by the Senate was chaired by Donald Stewart (D., Salem). Salem County already has three nuclear power plants operating or under construction—Stewart's support for S.C.S.1300 may well have been predicated on the exclusion of most of his district from any new hazardous waste facilities.

criteria are mandated by the bill, one following the announcement of preliminary criteria and the other subsequent to the issuance of revised criteria. The DEP is required by S.C.S.1300 to evaluate the comments made at these public hearings and then establish final siting criteria; this is to be accomplished within one year of the bill's enactment (that is, by September 1982). DEP is to establish these criteria in concert with the advisory council.

The role of the advisory council is to provide for additional public input at various stages of decisionmaking. The council consists of thirteen members appointed by the governor. Like the siting commission, the members of the advisory council are to represent diverse groups.

> Of these members, three shall be appointed from persons recommended by recognized environmental or public interest organizations; two from persons recommended by recognized organizations of municipal elected and appointed officials; two from persons recommended by recognized organizations of county elected and appointed officials; one from persons recommended by recognized community organizations; one from persons recommended by recognized organizations of firefighters; one from persons recommended by recognized organizations of industries which utilize onsite facilities for the treatment, storage or disposal of hazardous waste; one from persons recommended by recognized organizations of industries which utilize major hazardous waste facilities for the treatment, storage or disposal of hazardous waste; one from persons recommended by recognized organizations of persons licensed by the department to transport hazardous waste, or by individual licensed hazardous waste transporters; and one from persons recommended by recognized organizations or persons licensed by the department to operate major hazardous waste facilities, or by individual licensed major hazardous waste facility operators.[4]

One of the primary tasks of the siting commission is to apply DEP's siting criteria to designate actual sites for future waste disposal operations. The siting commission bases the number and type of sites to be needed on inventories and appraisals of the state's waste stream. These estimates are to be set forth in a Major Hazardous Waste Facilities Plan to be prepared by the siting commission. According to the bill, establishment of this plan is to occur during the same time that DEP is establishing siting criteria. Following the issuance of a draft plan, the siting commission is required to hold public hearings throughout the state prior to adopting a final plan—again, within one year of the bill's enactment.[b]

Once sites are designated by the siting commission, the municipalities selected are given six months to perform their own site suitability study (to be funded by the siting commission). A state administrative law judge then conducts an adjudicatory hearing on each proposed site. The affected municipality is a "party of interest" at this hearing, and may present testimony and cross-examine witnesses. The judge may deem a site to be suitable only if he or she determines that a waste facility on the proposed location would not constitute a "substantial detriment to the public health, safety and welfare of the effected municipality."[6] After the administrative law judge renders this decision, the siting commission is allowed to make a final evaluation of the site, subject only to judicial review. The municipal study thus essentially provides an opportunity for the local community to participate on a more informed basis during the adjudicatory proceedings. The bill does not require the siting commission to respond directly to the municipal findings, although these would likely be a factor in its final decision. This site designation process therefore is essentially *preemptive* (see Chapter 4); although local parties can provide input, at no point can the community render a decision of its own on the proposed site designation.

When the developer submits an application proposing to construct a specific facility on one of the designated sites, this same basic sequence is repeated. The emphasis now, however, is on the detailed facility plans and on the acceptability of the applicant, rather than on the site itself. The municipality in which the facility would be sited is again given six months to conduct a review of the proposed facility and its operator (to be funded by the applicant). Meanwhile, the siting commission is responsible for preparing an environmental and health impact statement regarding the proposed facility. The commission does not rule formally on the facility, however—this is a re-

[b]Substantial delays already apparent in the appointment of the Hazardous Waste Siting Commission members, however, mean that establishment of the plan will undoubtedly take longer than specified in the statute. At the beginning of 1982, DEP was working on establishing draft siting criteria, but the siting commission's hazardous waste facilities plan—supposedly to be established concurrently—had not been initiated due to the absence of a siting commission. Newly elected Governor Thomas Kean had not yet appointed any of the siting commission's members.[5]

sponsibility of the state's environmental agency (DEP). Soon after the municipality completes its study, DEP tentatively rejects or approves the facility. In the case of a tentative approval, an adjudicatory hearing is again held before an administrative law judge (where the affected municipality is again a "party of interest"). Following this hearing, ultimate decisionmaking authority reverts to DEP, which may affirm or reject the judge's recommendations. As in the case of the original site selection by the siting commission, there is no formal local decision—DEP's final decision is subject only to judicial scrutiny.

LEGITIMACY OF THE NEW JERSEY SITING PROCESS

The most important component of this innovative state siting process is the creation of a means to bypass local opposition to establishment of waste disposal operations. The powers granted to the state's new Hazardous Waste Facilities Siting Commission are thus intended to "guarantee" a way of bypassing such opposition. State eminent domain power can be invoked, regardless of any local land-use planning provisions.[7]

Despite this enhanced state power over actual siting decisions, one of the principal strengths of the New Jersey legislation is in its attention to a balance between state and local interests. Although the original bill introduced in the state senate was an entirely preemptive measure that closely resembled the Chemical Manufacturers Association's "model" siting bill,[c] the revision of this legislation by the Senate Energy and Environment Committee resulted in inclusion of far greater public participation throughout the siting process. Those who are likely to be affected most adversely by the siting of a facility,

[c]Under the original bill, a Hazardous Waste Facilities Corporation would have had the power to prepare a plan that included identification of possible sites for facilities (based on DEP siting criteria), exercise eminent domain authority to secure property necessary for a hazardous waste facility, certify private developers wishing to build facilities, and construct and operate public facilities if private capacity had not been established within two years. Concern for public participation was only nominally incorporated in this siting legislation. The bill provided for public hearings at two different stages in the process (directing the corporation to "consider" the testimony), and allowed municipalities to file written objections with the corporation to contest facility applications. These objections would, nevertheless, "in no way alter or interfere with the powers and duties of the corporation."[8]

particularly the residents of a host community, are provided several avenues for participation under S.C.S.1300's siting procedures. Such involvement extends well beyond the opportunity merely to voice complaints at required public hearings on the siting criteria and the Major Hazardous Waste Facilities Plan. Local residents also enjoy representation—both permanent and temporary—on the siting commission itself. Perhaps the most important innovation from the local standpoint, however, is the provision of funds for municipal studies, both of the siting commission's site designations and of DEP's review of facility applications. The opportunity to conduct such studies is one of the best means by which local residents can examine the relative advantages and disadvantages of any proposal that pertains to hazardous waste disposal in their "backyard."

New Jersey's siting legislation thus combines ultimate siting power in the hands of state decisionmakers with substantial input by local parties in such decisions. Such participation is seen as necessary to gain local residents' acceptance of this process. In order for successful decisionmaking to occur on a topic as controversial as the siting of hazardous waste facilities, it is imperative that the parties involved view the process as legitimate. The concept of a "balanced, sequential, and timely" siting procedure (described in Chapter 4) aims at establishing such legitimacy.

Perhaps the most important question about the perceived legitimacy of the siting process in New Jersey is the degree to which local participation and local views will actually be incorporated in state decisionmaking by DEP and by the newly established Hazardous Waste Facilities Siting Commission. The local role in this siting process may be quite significant; or it may be only a token one. This will depend on the manner in which the legislation is actually implemented. The bill does not guarantee a balanced and sequential siting procedure; by the same token, it does not preclude one either. This ambiguity can be traced to the administrative discretion inherent in the bill: the two agencies that administer this siting process will determine the degree to which power is balanced and decisions are made sequentially. When DEP makes a decision concerning a developer's application for a facility, for example, it is unclear how important a role it will accord to the mandated municipal study of the application and of the developer. This municipal study will have far less importance if, during the six months in which the application is being subjected to local scrutiny, DEP concurrently arrives at its tentative decision. Such an action, as noted previously, would tend to place the agency in the position of defending its already formulated decision (the familiar "decide, announce, defend" sequence), rather than remaining open to new evidence and viewpoints introduced by informed local input.

On the other hand, because DEP does not have to announce its preliminary decision on an application until two months after the municipal study has been completed, it is possible that the local study will form an important basis for the department's final decision. The bill includes one crucial provision in this regard: if DEP grants tentative approval to the application despite municipal findings which run contrary to such approval, DEP must provide specific reasons for having rejected the local conclusions.[9] Thus, the siting process may well prove to be "balanced and sequential," provided it is implemented with sufficient sensitivity to local concerns.[d]

Although the New Jersey siting statute pays considerable attention to the balance between state and local interests in siting, it is nevertheless apparent that this process is far more preemptive than are the alternative types of override decisionmaking advocated in Chapter 4. The hallmark of a state override process is its grant of authority to the local government to say "yes" or "no" to a proposed hazardous waste facility. Under this approach, decisionmaking by state agencies occurs only after local residents have had a reasonable chance to express their concerns and to arrive at a formal local judgment in pursuit of their own best interests. The New Jersey siting procedure, when judged by these criteria, falls short of providing for a true local decision. Despite the bill's requirement that municipalities be given funds to review both site designations and facility applications—an important feature of a balanced and sequential override process—such studies are not enough by themselves. At best, they allow local residents to become sufficiently informed about the technical aspects of specific site or facility proposals to participate in the ensuing state siting decisions (such information is especially important when the local opponents argue before the administrative law judge). The performance of a local study could instead be a *prerequisite* to local discussions and local decisionmaking about all aspects of a siting proposal in their area.

The absence of a formal local decision in the New Jersey siting process suggests that community response to a siting proposal will

[d]The notion of a timely process discussed in Chapter 4 is well incorporated in the bill. Municipal studies, for example, must be completed within six months. In fact, the entire siting processs includes specific time allocations concerning inception of the next stage.

have to be expressed through the findings of the municipal studies, and through the testimony of local opponents at the required adjudicatory hearings. Unfortunately, to the extent that this occurs, the siting debate would tend to be confined to discourse about empirical findings and scentific uncertainty. A local study of a proposed facility, for example, will undoubtedly evaluate the risks presented by that facility's design. Yet such a study is likely to lack a discussion of what constitutes "acceptable risk" in the eyes of these residents, or of the aspirations of the community in terms of its future land use, or of concerns over community image. Nevertheless, these types of local values are likely to be a substantial factor in opposition to (or acceptance of) proposed waste management facilities.

As a result, because the New Jersey statute deprives local residents of an opportunity to make an actual decision on a site designation or on a facility proposal—instead relying on two municipal studies—certain issues are less likely to be addressed, even though they are of compelling significance from a local standpoint.[e] Robert Socolow's discussion of "failures of discourse" in environmental disputes is thus germane to the New Jersey siting process:

> The failure of technical studies to assist in the resolution of environmental controversies is part of a larger pattern of failures of discourse in problems that put major societal values at stake. Discussions of goals, of visions of the future, are enormously inhibited. Privately, goals will be talked about readily, as one discovers in even the most casual encounter with any of the participants. But the public debate is cloaked in a formality that excludes a large part of what people most care about.[10]

PARTICIPATION FOR WHOM?

While New Jersey's hazardous waste facility siting statute does allow for substantial input into siting decisions by certain local actors, this

[e]The Starr County, Texas, case, described earlier, has shown how even with technically exemplary site location and facility design, public acceptance is far from guaranteed. Although this proposed landfill posed fewer risks than its counterparts in virtually any area across the country, local residents still opposed its construction—largely on the grounds of political equity. Unfortunately, had the Starr County landfill been reviewed under New Jersey's procedures, participatory measures featuring local studies of technical suitability would have largely obscured the *true* basis for local resistance. Community residents would quite rightly have felt frustrated when the siting debate focused on issues that were not closely related to their basic concerns.

"vertical" balance between state and local interests is no guarantee of full public participation. Also important is the extent to which various "horizontal" and "structural" interests are able to have a voice in siting (see Chapter 4). The drawback of the New Jersey legislation in this regard is its extreme deference to organized interests, and to the belief that elected local representatives will be able to act as true delegates for their constituents' viewpoints.

Reliance on participation by representatives during the siting of hazardous waste facilities carries with it the danger that many citizens—especially residents of the host community—will feel that they have been left out of the decisionmaking process, that their interests have not been adequately considered. Indeed, since residents of a municipality should not be expected to react uniformly to any given siting proposal, even the most sincere efforts by a local representative on behalf of his or her constituents will not provide complete representation of local interests. An obvious example of these potentially divergent interests is the difference in viewpoint between those who would live close to a proposed facility and are thus most exposed to its risks (abutters), and those who would receive direct financial benefits from the facility's construction and operation (e.g., future employees). These distinctions were apparent in Jackson Township, New Jersey, as shown in Appendix A.

Most of the opportunities for local participation in the New Jersey siting process are necessarily dominated by municipal representatives (the mayor, members of the town council, and so on). Such elected officials are included in the membership of the commission and advisory council. They supervise the studies of proposed sites and of facility designs. And they appear at adjudicatory hearings on site designations and facility applications. The typical community resident, however, has only one formal means of participation: attendance at the required public hearings.[f] Thus, for all its attention to public participation, involvement in the state siting process is clearly focused on a few select parties who, it is assumed,

[f]Even worse, the bill does not mandate that public hearings take place at the two most critical points of the siting process: designation of sites by the siting commission and review of facility applications by DEP. It is at these junctures in the New Jersey process, however, that the prospect of actually hosting a hazardous waste facility will become most evident to community residents. While such hearings are not mandatory, however, it is likely that they will be held, not only because of DEP administrative procedures that exist apart from S.C.S.1300 but also because of the advisory council's ability to hold whatever hearings it deems necessary.[11]

will be able to represent the countless other citizens who are not formally included in the decisionmaking effort.

A basic strength of this legislation, however, is the fact that it pays close attention to the structural dimension of citizen participation by attempting to incorporate the views of several important interest groups. In creating a siting commission, for example, the bill explicitly recognizes three main interest groups. An even broader group representation is found in the membership of the advisory council.

Inclusion of these diverse interest groups in the planning process may well facilitate the successful siting of hazardous waste facilities in New Jersey. Since those parties who have a stake in the outcome of a siting decision are allowed a role in this decisionmaking, their acceptance of this process should follow (that is, they should judge it to be a legitimate exercise of authority). Unfortunately, the actual extent of involvement of some of these groups is limited by this bill. Community organizations, for example, are recognized only by representation on the advisory council. Therefore, the process may instead be seen as a "token" measure that does not allow certain groups a genuine voice in siting. This problem will be particularly evident to local interests who must rely on a few elected officials to present their viewpoint throughout the siting process.

There is another drawback to the participatory framework—its attention to organized interests may mean a neglect of weaker structural groups who also have a stake in siting decisions. Residents of towns adjacent to a community which hosts a facility, for example, tend not to be represented in traditional siting processes; the New Jersey structure is no exception. To the extent that such communities will be subject to risks associated with certain hazardous waste disposal operations, it seems unfair to deny them a role in assessing such facilities. The principal danger of excluding nonstructural groups is that the ultimate decisions arrived at by recognized interests will overlook the interests of many in the state. Widespread perception of siting procedures as biased or inequitable signals a process whose legitimacy is uncertain and whose participatory mechanisms deserve reexamination.

The evolution of New Jersey's siting legislation was one in which the notion of meaningful public participation in decisionmaking was incorporated to an ever-increasing extent. Nevertheless, despite the bill's prudent vertical balance between state and local interests and its explicit involvement of important structural groups, the statute neglects many actors whose participation in siting is extremely desirable, if not essential. Such discrepancies are most apparent at the local level where elected officials are called upon to join the siting commission and advisory council, conduct studies of site designations and facility applications, as well as argue the local perspective before adjudicatory hearings. These provisions leave a critical "participatory void" which community residents around the state may come to see as characteristic of a siting process which caters more to select state interest groups than to all interested parties in the state.

AN ADVERSARIAL PROCESS

One way to rectify this participatory void would be to allow individuals like the abutters to a proposed facility to become involved directly in the decisionmaking process. The manner in which the New Jersey statute dictates that decisions be rendered, however, militates against inclusion of diverse horizontal and structural groups. At the root of this problem is the fact that the bill establishes an adversarial approach to producing decisions. This adversarial quality is nowhere more evident than in the required adjudicatory hearings before an administrative law judge. In establishing this unique decisionmaking feature, the authors of New Jersey's siting legislation envisioned local opposition to establishment of any hazardous waste facility. Thus they created a procedure whereby representatives of these communities might argue their case against those in favor of such siting. Any increase in the number of parties involved in this type of adversarial process would have complicated the problem of resolving these clearly competing claims.

The approach to siting adopted in New Jersey is thus very different from the negotiated decisionmaking format advocated in Chapters 4 and 5. The goal of negotiation is to bring together those groups who have a stake in siting so as (hopefully) to arrive at *mutually acceptable* solutions. In contrast, New Jersey's siting legislation involves a limited number of individuals and groups who compete to have their initial position adopted in the final decision.

For local communities who face either the prospect of being designated as a hazardous waste site or of hosting a specific waste treatment, storage, or disposal facility, New Jersey's adversarial approach may present more of a "win-lose" proposition than would a negotiated process. Assuming, for example, that most citizens in a town oppose the designation of their community as a future disposal site, these residents will claim a victory if the administrative law judge or the siting commission decides that the town is not a suitable location. Conversely, if the town is selected as a site by the commission, residents will feel they have lost the siting battle.

To the extent that adversarial mechanisms offer these win-lose alternatives,[g] they must focus on providing for a "fair fight" so that the losers cannot legitimately complain that they were unfairly treated. Although the notion of the adversarial "fair fight" is well embedded in American dispute resolution, it is not necessarily a desirable approach to such environmental controversies as the siting of a hazardous waste facility.[12] The losing side in the adversarial contest, for example, will employ all possible avenues to appeal and protest a decision which runs counter to their interests. In the context of the New Jersey siting process, this undoubtedly means that groups will initiate court suits to contest decisions by the siting commission or by the DEP. Indeed, the statute explicitly recognizes that these agencies' final decisions are subject to judicial review. The prospect of such legal battles, however, will often prove an insurmountable obstacle for facility developers who cannot afford the price of sustained delay (see Chapter 4). This problem of potential court action adds to what is already an acute problem—the long amount of time required to complete the various siting stages.[h]

[g]This problem is compounded by the basic emphasis of the New Jersey statute on the *process* of siting. As a result, the bill is primarily concerned with balancing the competing interests of various individuals and groups at the local and statewide level. A fundamental drawback to this legislation, therefore, is its scant attention to the *results* of siting. Among these results, of course, are risks to human health and the environment, the economic effects of facility construction and operation, the compatibility of a facility with existing land uses, and overwhelming local concerns about exploitation and equity. In Chapter 5, we advocate a negotiated process to address these impacts and thereby move instead toward a "win-win" approach to siting.

[h]From the time of the bill's enactment in 1981 to the first approval of a waste treatment, storage, or disposal facility will require between three and four years, even *without* judicial review. Thus, one of the problems with New Jersey's siting bill is its incentive to developers to bypass entirely the cumbersome siting framework it establishes. In fact, this was attempted in the first months after the statute's enactment by use of a "loophole" in the bill which allows for expansion of a facility's disposal capacity without reference to the bill's siting procedures, provided such expansion did not exceed 50 percent. The consequences of this provision are quite unfavorable, according to Peter Montague of the Environmental Research Foundation and Princeton University, since this encourages a developer to site a less controversial facility first (e.g., a waste storage station) and later add a more controversial operation (e.g., an incinerator) without following the explicit new process. In early 1982, DEP and the New Jersey Attorney General's office were reexamining this provision; DEP may well opt to limit this exemption so as to adhere more closely to the "spirit" of the siting bill.[13]

The win-lose aspect of New Jersey's adversarial approach to siting also means that extraordinary political pressures will impinge on those who have the power to influence the outcome of specific siting decisions. A potential weakness of the bill in this regard is its provision for possible gubernatorial veto of any siting commission action.[14] The notion behind creating a siting commission was that an independent body would best be able to handle the difficult task of selecting sites for future hazardous waste facilities; the independence of this agency presumably would insulate it from extreme political pressures which might be brought to bear on elected representatives or existing executive agencies. The possibility of a governor's veto in New Jersey means, however, that after the siting commission has finally made its site designations, communities may be able to derail the siting attempt by using their influence directly with the governor. Such political pressures—often successful—are not at all uncommon in siting attempts, as evidenced in Massachusetts where three sites that had been selected in a study as suitable for waste disposal facilities were quickly exempted by the state legislature.[15]

SUMMARY

New Jersey's siting statute wisely recognizes that public involvement and a tightly constrained state siting presence will be necessary to gain acceptance for establishment of hazardous waste storage, treatment, or disposal operations. Unfortunately, adversarial decision-making under the siting process may intensify the inevitable conflict and contentiousness of siting attempts, rather than channeling public involvement into more constructive avenues whose goal is to provide for mutually acceptable solutions.[i]

In considering New Jersey's "solutions" for the siting of hazardous waste facilities, it is important to keep in mind the climate of extreme opposition that typically prevails in such siting attempts (see Chapter 2). In the face of intense opposition, creation of a preemptive, adversarial siting process in New Jersey means that the state is likely to continue to experience the same types of problems that have beset unsuccessful facility siting attempts in the past.

[i]Use of negotiation, backed by state arbitration, is the best means to achieve this end, as argued in Chapters 4 and 5.

NOTES

1. New Jersey, P.L. 1981, Chapter 279, "Major Hazardous Waste Facilities Siting Act," Senate Committee Substitute for 1980 Senate No. 1300, Sec. 2.
2. Matthew Purdy, "Accord Set on Panel For Waste," *Trenton Times* (18 December 1980): B6.
3. New Jersey, Hazardous Waste Advisory Commission, *Report of the Hazardous Waste Advisory Commission to Governor Brendan Byrne* (Trenton: State of New Jersey, 1980), pp. 1–4; Diane Graves, "Hazardous Waste Bill Emerges from the Rubble," *Jersey Sierran* 8 (September-October 1980): 3; Tom Johnson, "Siting Criteria Backed for Toxic Waste Plants," *Newark Star Ledger* (26 June 1981): 30.
4. New Jersey, "Major Hazardous Waste Facilities Siting Act," Sec. 6(a).
5. Richard Gimello, Chief, Office of Public Participation, New Jersey Department of Environmental Protection, Interview, 17 February 1982.
6. New Jersey, "Major Hazardous Waste Facilities Siting Act," Sec. 11(4).
7. Ibid., Sec. 15.
8. New Jersey, "Hazardous Waste Facilities Corporation Act," Senate No. 1300, Sec. 22(b).
9. New Jersey, "Major Hazardous Waste Facilities Siting Act," Sec. 12(6).
10. Robert Socolow, "Failures of Discourse: Obstacles to the Integration of Environmental Values into Natural Resource Policy," in *When Values Conflict: Essays on Environmental Analysis, Discourse, and Decision* (Cambridge, Mass.: Ballinger Publishing Co., 1976), p. 2.
11. Gimello, Interview, 17 February 1982.
12. See, for example, Lawrence Susskind and Alan Weinstein, "Towards a Theory of Environmental Dispute Resolution," *Boston College Environmental Affairs Law Review* 9 (1980–81): 311–357.
13. Gimello, Interview, 17 February 1982; "Environmental Researcher Cites Loophole in New Jersey Facility Siting Law," *Hazardous Materials Intelligence Report* (23 October 1981): 10–11.
14. New Jersey, "Major Hazardous Waste Facilities Siting Act," Sec. 4(j). Such a provision for executive check on the commission's actions is required under New Jersey law. Gimello, Interview, 17 February 1982.
15. See chapter 4, footnote 26, *supra*.

BIBLIOGRAPHY

Ackerman, Bruce. "The Jurisprudence of Just Compensation." *Environmental Law* 7 (Spring 1977): 509–518.

Adelung, Warren J. et al. vs. Township of Jackson, New Jersey. U.S. District Court Civil 79; 2613.

Ahern, William. "California Meets the LNG Terminal." *Coastal Zone Management Journal* 7, no. 2-3-4 (1980): 216–221.

American Association for the Advancement of Science. *Scientific and Technical Aspects of Hazardous Waste Management*. Report from a workshop considering problems identified by the Intergovernmental Science, Engineering, and Technology Advisory Panel. Washington, D.C.: July 11–13, 1979.

American Law Institute. *A Model Land Development Code: Proposed Official Draft—Complete Text and Commentary*. Philadelphia: ALI, 15 April 1975.

Arbuckle, J. Gordon. "The Deepwater Port Act and Energy Facilities Siting: Hopeful Solution or Another Part of the Problem." *Natural Resources Lawyer* 9, no. 3 (1976): 511–516.

Arnstein, Sherry. "A Ladder of Citizen Participation." *The Politics of Technology*. Edited by Godfrey Boyle, David Elliott and Robin Roy. New York: Longman, Inc., 1977.

Ashford, Nicholas. "The Limits of Cost-Benefit Analysis in Regulatory Decisions." *Technology Review* 82 (May 1980): 70–72.

Bacow, Lawrence. *Mitigation, Compensation, Incentives and Preemption*. Prepared for the National Governors' Association, 10 November 1980.

Baram, Michael. *Environmental Law and the Siting of Facilities: Issues in Land Use and Coastal Zone Management.* Cambridge, Mass.: Ballinger Publishing Co., 1976.

Booz, Allen & Hamilton. *Hazardous Waste Management Capacity Development in the Delaware River Basin and New Jersey: A Program Strategy.* Prepared for the Delaware River Basin Commission and the New Jersey Department of Environmental Protection. Bethesda, Md.: 8 April 1980.

______. *Hazardous Waste Management Capacity Development in the State of New Jersey.* Prepared for the State of New Jersey and the Delaware River Basin Commission, Bethesda, Md.: 15 April 1980.

Bosselman, Fred and David Callies. *The Quiet Revolution in Land Use Control: Summary Report.* Prepared for Council on Environmental Quality. Washington, D.C.: U.S. Government Printing Office, 1971.

Brooks, Harvey. "Science and Trans-Science." *Minerva* 10 (1972): 484–486.

Brown, Michael. "Drums of Death." *Audubon* 82 (July 1980): 120–133.

______. *Laying Waste: The Poisoning of America by Toxic Chemicals.* New York: Pantheon Books, 1979.

______. "New Jersey Cleans Up Its Pollution Act." *New York Times Magazine,* 23 November 1980, pp. 142–146.

Burke, Edmund. *A Participatory Approach to Urban Planning.* New York: Human Sciences Press, 1979.

Byrd, J.F. "An Industrial Approach to Siting of Hazardous Waste Disposal Facilities." Speech to the National Conference on Management of Uncontrolled Hazardous Waste Sites, Washington, D.C.: 15–17 October 1980.

California. Auditor General. *Report By the Auditor General of California—California's Hazardous Waste Management Program Does Not Fully Protect the Public from the Harmful Effects of Hazardous Waste.* Sacramento: Auditor General, 26 October 1981.

California. Department of Health Services. *Waste Not, Want Not.* A Report of the Advisory Committee on Hazardous Waste Facility Siting Criteria. Sacramento: Governor's Office of Planning and Research, 30 June 1981.

California. Hearing: "On the Matter of Solving the Hazardous Waste Problem: Non-Toxic Solutions for the 1980's." Los Angeles, 17 November 1980.

California. Office of Appropriate Technology. *Alternatives to the Land Disposal of Hazardous Wastes: An Assessment for California.* Sacramento: OAT, 1981.

Cartwright, Keros, Robert Gilkeson, Robert Griffin, Thomas Johnson, David Lindorff, and Robert DuMontelle. *Hydrogeologic Considera-*

tions in Hazardous Waste Disposal in Illinois. Champaign: Illinois Geological Survey, Environmental Geology Notes 94, February 1981.

Casper, Barry and Paul David Wellstone. *Powerline: The First Battle of America's Energy War.* Amherst: University of Massachusetts Press, 1981.

Chemical Manufacturers Association. "A Statute for the Siting, Construction and Financing of Hazardous Waste Treatment, Disposal and Storage Facilities." Washington, D.C.: CMA, 1980.

Clark-McGlennon Associates. *Criteria for Evaluating Sites for Hazardous Waste Management.* Prepared for New England Regional Commission. Boston, Mass.: November 1980.

______. *A Decision Guide for Siting Acceptable Hazardous Waste Facilities in New England.* Prepared for New England Regional Commission. Boston, Mass.: November 1980.

______. *Institutional Arrangements for Developing Hazardous Waste Facilities in New England.* Prepared for New England Regional Commission. Boston, Mass.: 8 July 1980.

______. *An Introduction to Facilities for Hazardous Waste Management.* Prepared for New England Regional Commission. Boston, Mass.: November 1980.

______. *Negotiating to Protect Your Interests: A Handbook on Siting Acceptable Hazardous Waste Facilities in New England.* Prepared for New England Regional Commission. Boston, Mass.: November 1980.

Commonwealth of Massachusets. *Massachusetts Hazardous Waste Facility Siting Act.* Chapter 21D, Approved 15 July 1980.

Connecticut Hazardous Waste Siting Board. *Final Report of the Interim Study Committee,* January 1981.

Cook, Tom and James Knudson. *A History of Efforts to Acquire a Hazardous Waste Site in the State of Washington.* Presented to the Natural Resources Committee Meeting of the National Conference of State Legislatures. Denver, Colorado, 15–16 February 1980.

Corbett, Thomas. *Cancer and Chemicals.* Chicago: Nelson-Hall, 1979.

Cornaby, Barney. *Management of Toxic Substances in Our Ecosystems.* Woburn, Mass.: Ann Arbor Science, 1981.

Costle, Douglas and Eckhardt Beck. "Attack on Hazardous Waste: Turning Back the Toxic Tide." *Capital University Law Review* 9 (Spring 1980): 425–433.

Council on Environmental Quality et al. *Environmental Trends.* Washington, D.C.: U.S. Government Printing Office, 1981.

______. *Public Opinion on Environmental Issues: Results of a National Public Opinion Survey.* Washington, D.C.: U.S. Government Printing Office, 1980.

______ . "Toxic Substances and Environmental Health." *Environmental Quality: The Tenth Annual Report.* Washington, D.C.: U.S. Government Printing Office, 1979.

Dallaire, Gene. "Toxics in the N.J. Environment: Microcosm of U.S. Ills." *Civil Engineering* (September 1979): 74–86.

Deal, David. "The Durham Controversy: Energy Facility Siting and the Land Use Planning and Control Process." *Natural Resources Lawyer* 8, no. 3 (1975): 445–453.

Delaware River Basin Commission/New Jersey Department of Environmental Protection. Public Meeting on Level III Criteria for Identification and Screening of Sites for Hazardous Waste Facilities, Edison Township, New Jersey, 16 July 1980.

Ducsik, Dennis. *Electricity Planning and the Environment.* Cambridge, Mass.: Ballinger, forthcoming.

Egginton, Joyce. *The Poisoning of Michigan.* New York: Norton, 1980.

Ember, Lois. "Needed: Hazardous Waste Disposal (But Not In My Backyard)." *Environmental Science & Technology* 13 (August 1979): 913–915.

Environmental Defense Fund. *Malignant Neglect.* New York: Vintage, 1980.

Environmental Resources Management, Inc. *Technical Criteria for Identification and Screening of Sites for Hazardous Waste Facilities.* Prepared for the Delaware River Basin Commission and the New Jersey Department of Environmental Protection. West Chester, Penn.: August 1981.

Epstein, Samuel. *The Politics of Cancer.* New York: Anchor, 1979.

Ervin, David and James Fitch. "Evaluating Alternative Compensation and Recapture Techniques for Expanded Public Control of Land Use." *Natural Resources Journal* 19 (January 1979): 21–41.

Farkas, Alan. "Overcoming Public Opposition to the Establishment of New Hazardous Waste Disposal Sites." *Capital University Law Review* 9 (Spring 1980): 451–465.

Florida. *Florida Resource Recovery and Management Act.* HB 311, Chapter 80-302, Approved 2 July 1980.

Getz, Malcolm and Benjamin Walter. "Environmental Policy and Competitive Structure: Implications of the Hazardous Waste Management Program." *Policy Studies Journal* 9 (Winter 1980): 404–412.

Goetze, David. "A Decentralized Mechanism for Siting Hazardous Waste Disposal Facilities." Washington, D.C.: Resources for the Future, n.d. (Mimeo).

Goldfarb, William. "The Hazards of Our Hazardous Waste Policy." *Natural Resources Journal* 19 (April 1979): 249–260.

Golob, Richard, Adam Finkel and Robert Kunzig. *Hazardous Waste Management: The U.S. Perspective.* Cambridge, Mass.: World Information Systems, 1981.

Godwin, R. Kenneth and W. Bruce Shepard. "State Land Use Policies: Winners and Losers." *Environmental Law* 5 (Spring 1975): 703–726.

Graves, Diane. "Graves Resigns from Waste Panel." *Jersey Sierran* 8 (September-October 1980): 12.

______ . "Hazardous Waste Bill Emerges from the Rubble." *Jersey Sierran* 8 (September-October 1980): 3.

Greenwood, Richard. "Energy Facility Siting in North Dakota." *North Dakota Law Review* 52 (Summer 1976): 719–728.

Hall, Lawrence. "A Plague of Poisons." *National Wildlife* 17 (November/December 1979): 29–32.

Hanrahan, David. "Hazardous Wastes: Current Problems and Near-Term Solutions." *Technology Review* 82 (November 1979): 21–31.

Haskins, Peggy. "New Jersey's Hazardous Wastes." *New Jersey Voter* 51 (October 1980): 2, 4.

Healy, Robert and John Rosenberg. *Land Use and the States,* 2nd ed. Baltimore, Md.: Johns Hopkins University Press, 1979.

Hildyard, Nicholas. "Down in the Dumps." *The Ecologist* 9 (December 1979): 329–337.

Hoffman, Ian and William Gray. *A Study of the Jackson Township Sanitary Landfill.* Princeton: Princeton University Water Resources Program, 1980.

Institute of Environmental Research. *Facilities for Storage of PCB Waste: Selection of Sites and Handling Systems.* Prepared for Ontario Ministry of the Environment. Toronto: IER, April 1979.

Janerich, Dwight, William Burnett, Gerald Feck, Margaret Hoff, Philip Nasca, Anthony Polednak, Peter Greenwald, Nicholas Vianna. "Cancer Incidence in the Love Canal Area." *Science* 212 (19 June 1981): 1404–1407.

Kamlet, Kenneth. *Toxic Substances Programs in U.S. States and Territories: How Well Do They Work?* Washington, D.C.: National Wildlife Federation, 1979.

Keeney, Ralph. *Siting Energy Facilities.* New York: Academic Press, 1980.

Keystone Center. *Siting NonRadioactive Hazardous Waste Management Facilities: An Overview.* Final report of the First Keystone Workshop on Managing NonRadioactive Hazardous Wastes, Keystone, Colorado, September 1980.

______ . *Siting NonRadioactive Hazardous Waste Management Facilities—A Second Look.* Final Report of the Second Keystone Workshop on Siting NonRadioactive Hazardous Waste Management Facilities. Keystone, Colorado: Keystone Center, August 1981.

Kirkpatrick, Randall. "Toxic Waste in Your Own Backyard: Warning." *Central Jersey Monthly* 2 (December 1980): 34–37.

Kneese, Allen. "Environmental Policy." *The United States in the 1980s.* Edited by Peter Duignan and Alvin Rabushka. Palo Alto, California: Hoover Institution, Stanford University, 1980.

Lee, Craig. "Get the Public Involved in Planning." *Water & Wastes Engineering* (January 1980): 36–38.

Lowrence, William. *Of Acceptable Risk: Science and the Determination of Safety.* Los Altos, Cal.: William Kaufmann, Inc., 1976.

Luke, Ronald. "Managing Community Acceptance of Major Industrial Projects." *Coastal Zone Management Journal* 7, no. 2-3-4 (1980): 272–293.

McAvoy, James. "Hazardous Waste Management in Ohio: The Problem of Siting." *Capital University Law Review* 9 (Spring 1980): 450.

McMahon, Robert, Cindy Ernst, Ray Miyares, and Curtis Haymore. *Using Compensation and Incentives When Siting Hazardous Waste Management Facilities—A Handbook,* Washington, D.C.: U.S. EPA, 1982.

Magnuson, Ed. "The Poisoning of America." *Time,* 22 September 1980, pp. 58–66.

Maugh, Thomas. "Toxic Waste Disposal a Growing Problem." *Science* 204 (25 May 1979):819–823.

Michelman, Frank. "Property, Utility and Fairness: Comments on the Ethical Foundations of 'Just Compensation' Law." *Harvard Law Review* 80 (April 1967): 1165–1258.

Michigan. *Hazardous Waste Management Act.* Act 64, Effective 1 January 1980.

Montague, Peter. *Four Secure Landfills in New Jersey—A Study of the State of the Art in Shallow Burial Waste Disposal Technology.* Princeton, N.J.: Princeton University, Department of Chemical Engineering and Center for Energy and Environmental Studies, draft 1982.

Morell, David and Grace Singer. *Alternative Energy Facility Siting Policies for Urban Coastal Areas: Executive Summary of Findings and Policy Recommendations.* Prepared for U.S. Department of Energy, November 1980.

______. *Refining the Waterfront: Alternative Energy Facility Siting Policies for Urban Coastal Areas.* Cambridge, Mass.: Oelgeschlager, Gunn & Hain, Publishers Inc., 1980.

______. *State Legislatures and Energy Policy in the Northeast: Energy Facility Siting and Legislative Action.* Upton, N.Y.: Brookhaven National Laboratory, June 1977.

Mulvey, John. "An Incentive-Based Resource Recovery System: Reducing Improper Disposal of Hazardous Wastes." Princeton, N.J.: School of Engineering and Applied Science, Princeton University, March 1981.

Murray, William and Carl Seneker. "Implementation of an Industrial Siting Plan." *Hastings Law Journal* 3 (May 1980): 1073–1089.

______. "Industrial Siting: Allocating the Burden of Pollution." *Hastings Law Journal* 30 (November 1978): 301–336.

National Governors Association, *Review of Sixteen State Siting Laws.* Washington, D.C.: NGA, 1980.

______. Energy and Natural Resources Program. *Siting Hazardous Waste Facilities.* Final report of the National Governors' Association Subcommittee on the Environment. Washington, D.C.: National Governors' Association, March 1981.

National Wildlife Federation. *The Toxic Substances Dilemma: A Plan for Citizen Action.* Washington, D.C.: U.S. EPA, 1981.

New Jersey. Department of Environment Protection. *Abandoned Site Cleanup Status Report.* 20 November 1980.

New Jersey. Department of Health. Division of Epidemiology and Disease Control. *Groundwater Contamination and Possible Health Effects in Jackson Township, New Jersey.* Trenton: New Jersey Department of Health, 1980.

New Jersey. Hazardous Waste Advisory Commission. *Report of the Hazardous Waste Advisory Commission to Governor Brendan Byrne.* Trenton, N.J.: State of New Jersey, 1980.

New Jersey. *Major Hazardous Waste Facilities Siting Act.* S.C.S.-1300, Senate Committee Substitute, Adopted 13 January 1981. Signed as law 9 September 1981 (Chapter 279, Public Law of 1981).

New Jersey. Senate Energy and Environment Committee. *Public Hearing on S-1300: Volume I,* Trenton, N.J., 27 October 1980.

______. *Public Hearing on S-1300: Volume II.* Newark, N.J., 6 November 1980.

______. *Public Hearing on S-1300: Volume III.* Trenton, N.J., 3 December 1980.

O'Hare, Michael. "Enforcement and Source Reduction: Must We Subsidize Hazardous Waste?", Paper presented at National Conference on Hazardous Waste. Newark, N.J.: June 1980.

______. "Information Management and Public Choice." *Research in Public Policy Analysis and Management* (1): 223–256.

______. " 'Not On *My* Block You Don't': Facility Siting and the Strategic Importance of Compensation." *Public Policy* 25 (Fall 1977): 407–458.

______, Debra Sanderson and Lawrence Bacow. *Facility Siting.* New York: Van Nostrand-Reinhold, forthcoming. Draft manuscript.

Olpin, Owen. "Policing Toxic Chemicals." *Utah Law Review* 1 (1976): 85–94.

Olson, Mancur. *The Logic of Collective Action.* Cambridge, Mass.: Harvard University Press, 1971.

Peirce, J. Jeffrey and P. Aarne Vesilind. *Hazardous Waste Management.* Woburn, Mass.: Ann Arbor Science, 1981.

Pennsylvania. *Solid Waste Management Act.* HB 1840, 1979 Session, Approved 7 July 1980.

Peters, J. Douglas. "Durham, New Hampshire: A Victory for Home Rule?" *Ecology Law Quarterly* 5, no. 1 (1975): 63–67.

Pojasek, Robert. "Developing Solutions to Hazardous Waste Problems." *Environmental Science and Technology* 14 (August 1980): 924–929.

______. "Disposing of Hazardous Chemical Wastes." *Environmental Science & Technology* 13 (July 1979): 810–814.

______. ed. *Toxic and Hazardous Waste Disposal.* Woburn, Mass.: Ann Arbor Science, 1979 (vols. 1 and 2), 1980 (vols. 3 and 4), 1982 (vols. 5 and 6).

RL Associates. *Illegal Disposal of Hazardous Wastes: A Survey of Licensed Haulers.* Conducted for East Windsor Township, New Jersey. Princeton, N.J.: R.L. Associates, September 1980.

Raloff, Janet. "Abandoned Dumps: A Chemical Legacy." *Science News,* 26 May 1979, pp. 348–351.

Sasnett, Sam. *A Toxics Primer.* Washington, D.C.: League of Women Voters Education Fund, 1979.

Schiefelbein, Susan. "The Perils of Chemical Waste." *Saturday Review,* 5 January 1980, pp. 10–12.

Schroth, Peter. "Public Participation in Environmental Decisionmaking: A Comparative Perspective." *The Forum* 14 (Fall 1978): 352–368.

Seltz-Petrash, Ann. "Siting Hazardous Waste Facilities: Major Problem of the '80s." *Civil Engineering* 51 (January 1981): 14.

Sheiman, Deborah. *A Hazardous Waste Primer.* Washington, D.C.: League of Women Voters Education Fund, 1980.

Singer, Grace. . . . . *Nor Any Drop to Drink: Public Policies Toward Chemical Contamination of Drinking Water.* Princeton, N.J.: Princeton University, Center for Energy and Environmental Studies. February 1982 (draft).

Slaski, Lawrence. "Facility Siting and Locational Conflict Resolution." *Coastal Zone '78,* volume 1. New York: American Society of Civil Engineers, 1978.

Smalley, R.D. "Risk Assessment: An Introduction and Critique." *Coastal Zone Management Journal* 7, no. 2-3-4 (1980): 137–161.

Standley, David and Anthony Cortese. "NERCOM Issue Paper: Site Selection Process." Reprinted in *Siting of Hazardous Waste Management Facilities and Public Opposition,* SW-809. Washington, D.C.: U.S. Environmental Protection Agency, November 1979.

Steeler, Jonathan. *A Legislator's Guide to Hazardous Waste Management.* Prepared for National Conference of State Legislatures. Denver, Colorado: National Conference of State Legislatures, 15 October 1980.

Susskind, Lawrence. *Citizen Participation in the Siting of Hazardous Waste Facilities: Options and Observations,* draft. Prepared for the National Governors' Association, November 1980.

______ and Stephen Cassella. "The Dangers of Preemptive Legislation: The Case of LNG Facility Siting in California." *Environmental Impact Assessment Review* 1 (1980): 9–26.

Susskind, Lawrence and Michael O'Hare. *Managing the Social and Economic Impacts of Energy Development.* Summary Report of Phase 1 of the MIT Energy Impacts Project, Laboratory of Architecture and Planning, December 1977.

Susskind, Lawrence and Alan Weinstein. "Towards a Theory of Environmental Dispute Resolution." *Boston College Environmental Affairs Law Review.* Vol. 9, no. 2 (1980–81): 311–357.

Thurow, Lester. *The Zero-Sum Society: Distribution and the Possibilities for Economic Change.* New York: Basic Books, Inc., 1980.

Trauberman, Jeffrey. "Compensation for Environmental Pollution: An Overview of Salient Legal and Practical Issues" *ALI-ABA Course of Study Materials: Toxic Substances and Hazardous Waste* (Washington, D.C.: Environmental Law Institute, 1980), pp. 302–324.

U.S. Congress. House. Committee on Interstate and Foreign Commerce. *Hazardous Waste Disposal.* Hearings before the Subcommittee on Oversight and Investigations, 96th Cong., 1st sess., 1979.

U.S. Environmental Protection Agency. *Hazardous Waste Facility Siting: A Critical Problem,* SW-865. Washington, D.C.: U.S. Environmental Protection Agency, July 1980.

______. Office of Public Awareness. *EPA Journal,* vol. 5, no. 2. Washington, D.C.: U.S. Environmental Protection Agency, February 1979.

______. Office of Toxic Substances. *State Administrative Models for Toxic Substances Management.* Washington, D.C.: U.S. EPA, July 1980.

______. Office of Water and Waste Management. *Everybody's Problem: Hazardous Waste,* SW-826. Washington, D.C.: U.S. Environmental Protection Agency, 1980.

______. Office of Water and Waste Management. *Hazardous Waste Generation and Commercial Hazardous Waste Management Capacity: An Assessment,* SW-894. Washington, D.C.: U.S. Environmental Protection Agency, December 1980.

______. Office of Water and Waste Management. *Hazardous Waste Information,* SW-737. Washington, D.C.: U.S. Environmental Protection Agency, June 1980.

______. Office of Water and Waste Management. *Siting of Hazardous Waste Management Facilities and Public Opposition,* SW-809. Washington, D.C.: U.S. Environmental Protection Agency, November 1979.

______ . Oil and Special Materials Control Division. *Damages and Threats Caused by Hazardous Material Sites.* Washington, D.C.: U.S. Environmental Protection Agency, May 1980.

Urban Systems Research and Engineering, Inc. *A Handbook for the States on the Use of Compensation and Incentives in the Siting of Hazardous Waste Management Facilities,* draft. Prepared for U.S. Environmental Protection Agency. Cambridge, Mass.: Urban Systems Research and Engineering, Inc., 30 September 1980.

Wei, Norman. "Problems plague hazardous waste disposal." *Water and Wastes Engineering* (February 1980): 54–56.

Weimar, Robert. "Prevent Goundwater Contamination Before It's Too Late." *Water and Wastes Engineering* (February 1980): 30–33, 63.

Weinberg, Alvin. "Science and Trans-Science." *Minerva* 10 (1972): 209–222.

Weinberg, David. "We Almost Lost Elizabeth." *New Jersey Monthly,* August 1980, pp. 35–39, 97–113.

INDEX

ABOUT THE AUTHORS

David Morell is a research political scientist at Princeton University's Center for Energy and Environmental Studies, and a lecturer in the Department of Politics. Morell received an M.P.A. in Public and International Affairs and a Ph.D. in Political Science and Public Policy from Princeton. Since 1974, he has directed the Center's research program on land use and facility siting and taught environmental politics and policies. Previously, with the U.S. Environmental Protection Agency in Washington, D.C., he was acting director, Office of Transportation and Land Use Policy, Air Programs; director of the Coordination Office for Air Programs; and national coordinator for municipal water permit programs. Morell has spoken frequently on siting of hazardous waste facilities to Congressional committees, state legislative hearings, national conferences, and similar groupings. In 1980–81 he was a consultant on this topic to the Office of Science and Technology Policy, Executive Office of the President, and he is a consultant on facility siting to the Office of Energy of the Port Authority of New York and New Jersey. Morell is the author of many publications including *Centralized Power: The Politics of Scale in Electricity Generation* (1979) and *Refining the Waterfront: Alternative Energy Facility Siting Policies for Urban Coastal Areas* (1980).

Christopher Magorian is a 1981 graduate of Princeton University's Woodrow Wilson School of Public and International Affairs. His thesis research focused on policies to overcome public opposition to the siting of hazardous waste facilities. While working as a student intern for the New Jersey Department of Environmental Protection, he monitored the evolution of the state's hazardous waste facility siting legislation (Senate Bill 1300). Since graduation, Magorian has worked for the Pennsylvania Environmental Research Foundation's project on educating the public to participate in the facility siting process. His principal interests are the political, economic, and legal aspects of environmental protection. Starting in September 1982, Magorian will be a J.D. candidate at Yale Law School.

DATE D

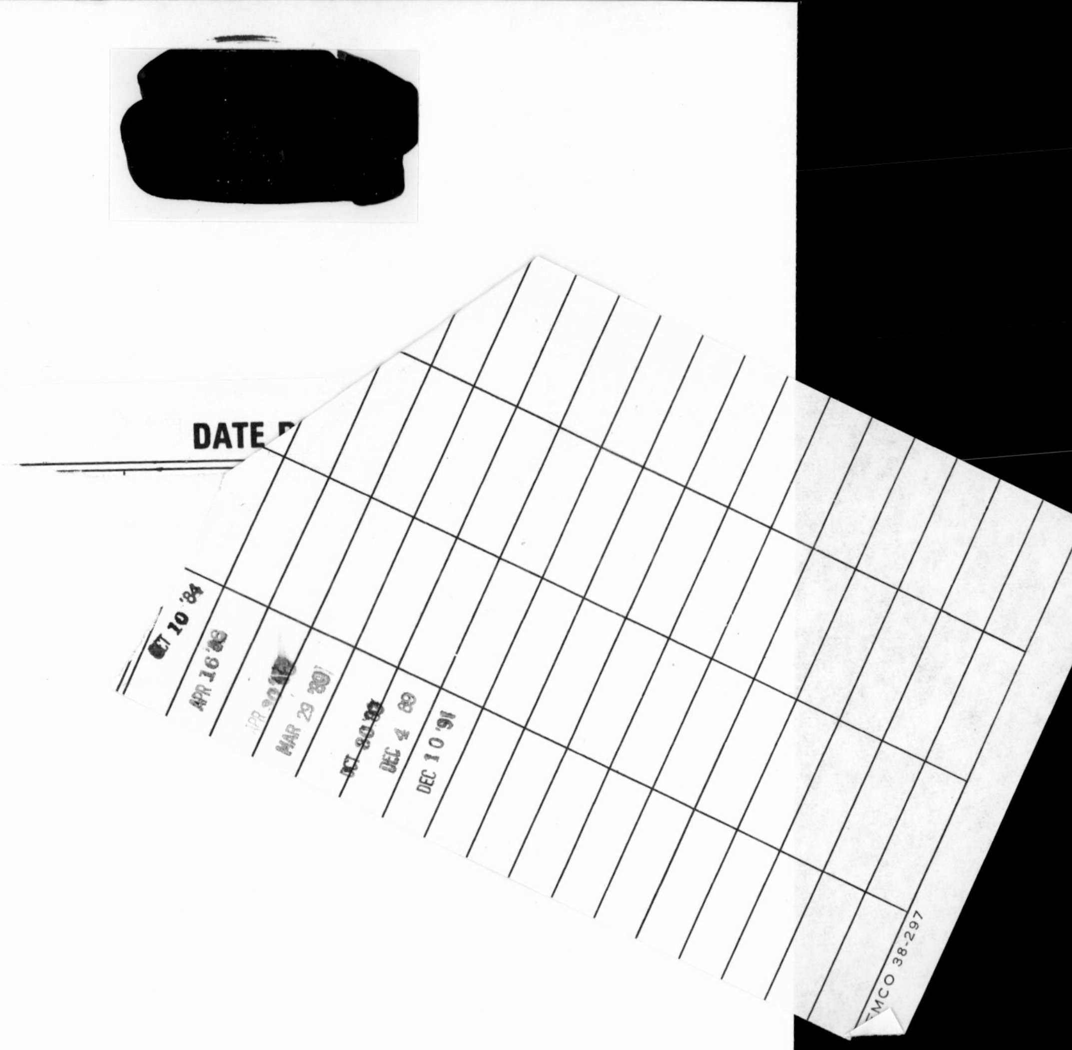